Planet Earth
ICE AGES

Other Publications:

MYSTERIES OF THE UNKNOWN
TIME FRAME
FIX IT YOURSELF
FITNESS, HEALTH & NUTRITION
SUCCESSFUL PARENTING
HEALTHY HOME COOKING
UNDERSTANDING COMPUTERS
LIBRARY OF NATIONS
THE ENCHANTED WORLD
THE KODAK LIBRARY OF CREATIVE PHOTOGRAPHY
GREAT MEALS IN MINUTES
THE CIVIL WAR
COLLECTOR'S LIBRARY OF THE CIVIL WAR
THE EPIC OF FLIGHT
THE GOOD COOK
WORLD WAR II
HOME REPAIR AND IMPROVEMENT
THE OLD WEST

For information on and a full description of any of the
Time-Life Books series listed above, please call
1-800-621-7026 or write:
  Reader Information
  Time-Life Customer Service
  P.O. Box C-32068
  Richmond, Virginia 23261-2068

This volume is one of a series that examines the
workings of the planet earth, from the geological
wonders of its continents to the marvels of its
atmosphere and its ocean depths.

Cover
A clump of arctic poppies blooms near the
edge of the icecap on the northern tip
of Ellesmere Island, Canada. Well adapted to a
bitterly cold environment, these hardy
flowers can produce seeds in less than a month.

Planet Earth

ICE AGES

By Windsor Chorlton
and The Editors of Time-Life Books

Time-Life Books, Alexandria, Virginia

PLANET EARTH

EDITOR: Thomas A. Lewis
Designer: Albert Sherman
Chief Researcher: Pat S. Good

Editorial Staff for *Ice Ages*
Associate Editors: Russell B. Adams Jr. (text);
Peggy Sawyer Seagrave (pictures)
Text Editors: Sarah Brash, Jan Leslie Cook
Staff Writers: Tim Appenzeller, William C. Banks,
Paul N. Mathless
Researchers: Megan Barnett and Barbara Moir
(principals), Jean Crawford, Melva Morgan
Holloman, Marilyn Murphy
Assistant Designer: Susan K. White
Copy Coordinators: Victoria Lee, Bobbie C. Paradise
Picture Coordinator: Donna Quaresima
Editorial Assistant: Caroline A. Boubin

Editorial Operations
Design: Ellen Robling (assistant director)
Copy Room: Diane Ullius
Production: Anne B. Landry (director), Celia Beattie
Quality Control: James J. Cox (director), Sally Collins
Library: Louise D. Forstall

Correspondents: Elisabeth Kraemer-Singh (Bonn);
Margot Hapgood, Dorothy Bacon (London); Miriam
Hsia, Lucy T. Voulgaris (New York); Maria Vincenza
Aloisi, Josephine du Brusle (Paris); Ann Natanson
(Rome). Valuable assistance was also provided by:
Helga Kohl (Bonn); Lois Lorimer (Copenhagen);
Robert W. Bone (Honolulu); Lesley Coleman,
Millicent Trowbridge (London); John Dunn
(Melbourne); Carolyn Chubet (New York); Dag
Christensen, Bent Onsager (Oslo); Mary Johnson
(Stockholm).

Library of Congress Cataloguing in Publication Data
Chorlton, Windsor.
   Ice ages.
   (Planet earth; 9)
   Bibliography: p.
   Includes index.
   1. Glacial epoch. I. Time-Life Books.   II. Title.
   III. Series.
QE697.C475 1983   551.7'92   82-16765
ISBN 0-8094-4328-7
ISBN 0-8094-4329-5 (lib. bdg.)

THE AUTHOR

Windsor Chorlton is a former Time-Life edi-
tor who now lives and works in London as a
freelance writer. He is a fellow of the Royal
Geographical Society and the author of *Cloud-
Dwellers of the Himalayas*, a volume in Time-Life
Books' Peoples of the Wild series.

THE CONSULTANT

Glaciologist Colin Bull is Professor of Geology and Mineralogy and Dean of the College
of Mathematical and Physical Sciences at Ohio
State University. He is a former director of the
university's Institute of Polar Studies and has led
several scientific expeditions to Antarctica.

CONTENTS

A LAND WHERE ICE STILL REIGNS

"A vast frozen sahara, immeasurable to the human eye" confronted the American explorer Isaac Hayes as he ascended the Greenland Ice Sheet in 1860. An expanse of ice almost continental in scale, the ice sheet is a deep-frozen reminder of times when Hayes's description would have applied to a third of the world's land area: the ice ages that have repeatedly gripped the planet over the past two million years.

Four million years ago, Greenland was largely free of ice. But something changed the earth's climatic balance, and, in the mountains that rim the island, winter snowfalls began to last through the summers without melting; as the snow accumulated from year to year, it compacted of its own weight until it flowed toward both the coast and the center of Greenland, eventually forming a blanket of ice up to two miles thick. Like the ice-age glaciers it preceded and outlived, the ice sheet buried the valleys and uplands beneath its immense weight, slowly depressing the island 2,000 feet or more into a saucer shape.

During the past 12,000 years, the margins of the ice sheet have receded an average of 120 miles. Relieved of the weight of ice, the newly exposed land has risen buoyantly toward its former level, just as parts of Europe and North America have been rebounding since the retreat of ice-age glaciers. But despite its recent waning, the Greenland Ice Sheet will probably survive to witness the next ice age, when trackless saharas of ice extend again to whiten much of the globe.

A dog sled crosses a glittering Arctic seascape of pack ice and trapped icebergs off the west coast of Greenland. Native Greenlanders rarely venture onto the lifeless inland ice sheet.

Hinting at the topography buried beneath the featureless interior ice, bluffs of volcanic rock mantled by recent snowfalls emerge near the eastern

edge of the Greenland Ice Sheet. Mountains in this part of Greenland reach heights of 6,500 feet and virtually cut off the ice sheet from the sea.

Cliffs bared by the retreat of the Greenland Ice Sheet since its maximum advance during the last ice age guard the coastline of York peninsula in

Muldrow Glacier, alight in a summer sunset,
retains its grip on Alaska's Mount McKinley—
perceptible evidence that the Pleistocene
ice epoch continues even today.

ACKNOWLEDGMENTS

For their help in the preparation of this book the editors wish to thank: In Bermuda: Ferry Reach—Dr. Thomas M. Iliffe, Bermuda Biological Station for Research. In Canada: Ottawa—Dr. Weston Blake, Geological Survey of Canada; Toronto—The Donner's Fund of Canada; University of Toronto; Vancouver—J. Ross Mackay. In Denmark: Copenhagen—Dr. Willi Dansgaard, Dr. Claus Hammer, Geophysical Isotope Laboratory, University of Copenhagen; Tony Higgens, The Geological Survey of Greenland. In France: Paris—André Leroi-Gourhan, Collège de France; Osmonde de Barante, Musée de l'Homme; Poitiers—Max Deynoux, Laboratoire de Pétrologie de la Surface, Université de Poitiers; Rueil-Malmaison—Bernard Biju-Duval, Olivier Gariel, Institut français du Pétrole. In Great Britain: Birmingham—Dr. G. R. Cope, University of Birmingham; Cambridge—Julian Paren, Eric W. Wolff, The British Antarctic Survey; Edinburgh—Commander Angus Erskine; Tom Bain, Institute of Geological Sciences; J. B. Sissons, University of Edinburgh; Leeds—Keith Thornton, Institute of Geological Sciences; Leyburn—Professor Cuchlaine King; London—Professor W. G. Chaloner, Bedford College; Martin Pulsford, Institute of Geological Sciences; Celina Fox, Museum of London; Sadie Alford, Novosti Press Agency; Miss C. Stott, The Observatory, Greenwich; Library, the Royal Society; Norfolk—Dr. Bernard Campbell; Martin Warren, The Cromer Museum; Oxfordshire—Dr. I.E.S. Edwards; Powys—Martin Farr; Sheffield—Dr. D. W. Humphries, University of Sheffield. In Ireland: Dublin—Dr. Jean Archer, Irish Geological Survey. In Italy: Florence—Mara Miniati, Istituto e Museo di Storia della Scienza; Genoa—Nicoletta Morello; Rome—Giorgio Buonvino, Osservatorio Astronomico. In the United States: Alaska—(Fairbanks) Dr. John Kelly, Institute of Marine Science, University of Alaska; (Gustavus) Bruce Paige, Gary Vequist, Glacier Bay National Monument; Arizona—(Tempe) Troy L. Péwé, Arizona State University; (Tucson) Agnes Paulson, Kitt Peak National Observatory; Julio L. Betancourt, Kenneth L. Cole, Douglas J. Donahue, C. W. Ferguson, Steven W. Leavitt, Paul S. Martin, Jim I. Mead, Robert Thompson, University of Arizona; California—(Los Angeles) William Akersten, Curator, Gregory P. Byrd, Superintendent, George C. Page Museum; Gretchen Sibley, Rancho La Brea; Dr. Clement Meighan, Anthropology Department, University of California, Los Angeles; (La Jolla) John Shelton; (San Diego) Dr. Stuart Hurlbert, San Diego State University; Colorado—(Boulder) Dr. Charles Barth, Dr. J. T. Hollin, University of Colorado; (Denver) Charles W. Naeser, U.S. Department of the Interior/Geological Survey; Delaware—(Newark) John F. Wehmiller, University of Delaware; District of Columbia—Dr. Roger Lewin; Winifred Reuning, National Science Foundation; Dr. Gary Haynes, Dr. Ian G. Macintyre, Dr. Clayton Ray, Dr. Dennis Stanford, Smithsonian Institution; Florida—(Coral Gables) Cesare Emiliani, University of Miami; (Gainesville) Dr. Neil D. Opdyke, University of Florida; (Miami Beach) Harold Hudson, Barbara Lidz, Jean Shinn, Fisher Island Station, U.S. Geological Survey; Georgia—(Savannah) Larry P. Atkinson, Skidaway Institute of Oceanography; Hawaii—(Hilo) Dr. Kinsell Coulson, Director, Thomas DeFoor, Mauna Loa Observatory, National Oceanic and Atmospheric Administration (NOAA); Illinois—(Chicago) J. Clay Bruner, Nina Cummings, Field Museum of Natural History; Christopher R. Scotese, University of Chicago; (Springfield) Dr. Russell Graham, Illinois State Museum; Maryland—(Annapolis) Rob Wood; (Bowie) I'Ann Blanchette; (College Park) Professor Anandu Vernakar, University of Maryland; (Greenbelt) Dr. Gerald North, Dr. H. Jay Zwally, Laboratory for Atmospheric Sciences, Goddard Space Flight Center, National Aeronautics and Space Administration (NASA); (Rockville) Dr. Robert Stuckenrath, Smithsonian Institution; (Suitland) Michael Matson, NOAA; (Temple Hills) Richard Legeckis, NOAA; Massachusetts—(Cambridge) Ann Blum, The Agassiz Museum, Harvard University; (Natick) John V. E. Hanson, U.S. Army Research and Development Laboratory; (Woods Hole) Robert N. Oldale, U.S. Geological Survey; K. O. Emery, Susumu Honjo, Woods Hole Oceanographic Institution; Michigan—(Detroit) Dr. Morris Goodman, School of Medicine, Wayne State University; Minnesota—(Minneapolis) Professor Edward P. Ney, University of Minnesota; Nebraska—(Lincoln) Bruce Koci; New Hampshire—(Hanover) Dr. Tony Gow, CRREL; New Jersey—(Princeton) Sheldon Judson, Princeton University; New Mexico—(Albuquerque) Rodman E. Snead, University of New Mexico; New York—(Amherst) Chester C. Langway Jr., State University of New York at Buffalo; (Ithaca) Dr. Arthur Bloom, Cornell University; (New York) Pamela Haas, American Museum of Natural History; Dr. Rhodes Fairbridge; (Palisades) Allan W. H. Bé, R. M. Cline, Rosemary Free, Frank Hall, Dr. J. D. Hays, Grace Irving, Gregory Kolibas, David Lazarus, Anne Lewis, Ms. Rusty Lotti, Dr. A. McIntyre, Alan Mix, Dr. William Ruddiman, Dr. Constance Sancetta, Sally Savage, Goesta Wollin, Lamont-Doherty Geological Laboratory; North Carolina—(Chapel Hill) Professor Conrad Neumann, The University of North Carolina at Chapel Hill; (Durham) David Burney, S. Duncan Heron Jr., Duke University; Ohio—(Columbus) Dr. Roger K. Burnard, Dr. Dwight DeLong, Lonnie G. Thompson, Dr. Peter Webb, Ian M. Whillans, Professor Sidney E. White, Ohio State University; (Worthington) Pei-Shing Wu; Oklahoma—(Tulsa) Robin G. Lighty, Cities Service Company; Pennsylvania—(Maytown) Ken Townsend; (Philadelphia) Henry N. Michael, Elizabeth K. Ralph, University of Pennsylvania; Rhode Island—(Providence) Dr. John Imbrie, Robley K. Matthews, Rosalind M. Mellor, Professor Thomas Webb III, Brown University; Texas—(Dallas) Russell S. Harmon, Southern Methodist University; Utah—(Salt Lake City) Dr. James Whelan, University of Utah; Virginia—(Alexandria) Walter Hilmers Jr.; (Arlington) Bill Hezlep; (Reston) Barbara Chappell, Carol Horan, Glen A. Izett, Jon Sellin, Trudy Sinnott, Robert Tilling, Henry Zoller, U.S. Geological Survey; Washington—(Bellingham) Maurice L. Schwartz, Coastal Consultants; (Seattle) Dr. F. C. Ugolini, College of Forest Resources; Dr. Lawrence C. Bliss, Donald Brownlee, Stephen C. Porter, Alan S. Thorndike, University of Washington; Wisconsin—(Madison) John E. Kutzbach, Kelly Redmond, University of Wisconsin; Wyoming—(Laramie) Dr. David J. Hofmann, University of Wyoming. In West Germany: Bonn—Dr. Eckhart Joachim, Rheinisches Landesmuseum; Hildesheim—Dr. Arne Eggebrecht, Römer und Pelizaeusmuseum; Koblenz—Dr. Horst Fehr, Landesamt für Denkmalpflege; Leipzig—Karen Stietzel, Urania-Verlag; Münster—Hermann-Josef Höper, Geologisches-Paläontologisches Institut und Museum der Universität Münster; Tübingen—Dr. Joachim Hahn, Institut für Urgeschichte, Universität Tübingen. In Yugoslavia: Belgrade—Serbian Academy of Sciences and Arts.

The editors also wish to thank the following persons: Pavle Svabic, Belgrade; Joanne Reid, Chicago; Dorothy Slater, Denver; Robert Kroon, Geneva; Lance Keyworth, Helsinki; Bob Schrepf, Lincoln, Nebraska; Cheryl Crooks, Los Angeles; Trini Bandres, Madrid; Cronin Buck Sleeper, Manchester, Vermont; David Hessekiel, Mexico City; Felix Rosenthal, Moscow; Juliet Tomlinson, Northampton, Massachusetts; Rowan Callick, Papua New Guinea; June Taboroff, Ann Wise, Rome.

Particularly useful sources of information and quotations used in this volume were: *Studies on Glaciers Preceded by the Discourse of Neuchâtel* by Louis Agassiz, translated and edited by Albert V. Carozzi, Hafner Publishing Co., 1967; *The Last Great Ice Sheets* edited by George H. Denton and Terence J. Hughes, John Wiley & Sons, 1981; *Glacial and Quaternary Geology* by Richard Foster Flint, John Wiley & Sons, 1971; *Ice Ages: Solving the Mystery* by John Imbrie and Katherine Palmer Imbrie, Enslow Publishers, 1979; *The Winters of the World* edited by Brian S. John, John Wiley & Sons, 1979; and *Pleistocene Extinctions: The Search for a Cause* edited by P. S. Martin and H. E. Wright Jr., Yale University Press, 1967.

The index was prepared by Gisela S. Knight.

BIBLIOGRAPHY

Books

Agassiz, Elizabeth Cary (ed.), *Louis Agassiz: His Life and Correspondence.* 2 vols. Houghton, Mifflin, 1885.

Agassiz, Louis, *Studies on Glaciers Preceded by the Discourse of Neuchâtel.* Transl. and ed. by Albert V. Carozzi. Hafner Publishing Co., 1967.

Anthes, Richard A., John J. Cahir, Alistair B. Fraser and Hans A. Panofsky, *The Atmosphere.* Charles E. Merrill, 1981.

Banks, Michael, *Greenland.* Rowman and Littlefield, 1975.

Bryson, Reid A., and Thomas J. Murray, *Climates of Hunger: Mankind and the World's Changing Weather.* University of Wisconsin Press, 1977.

Buckland, William, *Reliquiae Diluvianae; or Observations on the Organic Remains Contained in Caves, Fissures and Diluvial Gravel and on Other Geological Phenomena, Attesting the Action of an Universal Deluge.* London: John Murray, 1824.

Bulfinch, Thomas, *The Age of Fable, or Beauties of Mythology.* New American Library, 1962.

Calder, Nigel, *The Weather Machine.* London: British Broadcasting Corp., 1974.

Carrington, Richard, *Mermaids and Mastodons: A Book of Natural and Unnatural History.* Rinehart, 1957.

Chorley, Richard J., Antony J. Dunn and Robert P. Beckinsale, *The History of the Study of Landforms, or the Development of Geomorphology.* Methuen, 1964.

Claiborne, Robert, *Climate, Man and History.* Angus & Robertson, 1970.

Colbert, Edwin H., *Wandering Lands and Animals.* E. P. Dutton, 1973.

Coleman, A. P., *Ice Ages: Recent and Ancient.* AMS Press, 1969.

Croll, James, *Climate and Time in Their Geological Relations: A Theory of Secular Changes of the Earth's Climate.* London: Daldy, Isbister, & Co., 1875.

Czerkas, Sylvia Massey, and Donald F. Glut, *Dino-*

saurs, Mammoths, and Cavemen. E. P. Dutton, 1982.

Denton, George H., and Terence J. Hughes (eds.), *The Last Great Ice Sheets.* John Wiley & Sons, 1981.

Digby, Bassett, *The Mammoth and Mammoth-Hunting in North-East Siberia.* London: H. F. & G. Witherby, 1926.

Downie, C., and P. Wilkinson, *The Geology of Kilimanjaro.* Sheffield, England: University of Sheffield, 1972.

Eddy, John A., *A New Sun: The Solar Results from Skylab.* National Aeronautics and Space Administration, 1979.

Embleton, Clifford, and Cuchlaine A. M. King, *Periglacial Geomorphology.* John Wiley & Sons, 1975.

Escher, Arthur, and W. Stuart Watt (eds.), *Geology of Greenland.* Copenhagen: Geological Survey of Greenland, 1976.

Fairbridge, Rhodes W. (ed.), *The Encyclopedia of Geomorphology (Encyclopedia of Earth Sciences,* Vol. 3). Dowden, Hutchinson & Ross, 1968.

Ferguson, C. W., *Concepts and Techniques of Dendrochronology.* University of California Press, 1970.

Flint, Richard Foster, *Glacial and Quaternary Geology.* John Wiley & Sons, 1971.

Fodor, R. V., *Frozen Earth: Explaining the Ice Ages.* Enslow Publishers, 1981.

Forrester, Glenn C., *Niagara Falls and the Glacier.* Exposition Press, 1976.

Fraas, E., et al., *Jahreshefte des Vereins für Vaterländische Naturkunde in Württemberg.* Stuttgart: Klett & Hartmann, 1904.

Frison, George C., *Prehistoric Hunters of the High Plains.* Academic Press, 1978.

Frison, George C. (ed.), *The Casper Site: A Hell Gap Bison Kill on the High Plains.* Academic Press, 1974.

Fristrup, Børge, *The Greenland Ice Cap.* University of Washington Press, 1966.

Garašinin, Milutin (ed.), *La Vie et L'Oeuvre de Milutin Milanković, 1879-1979.* Belgrade: Serbian Academy of Sciences and Arts, 1982.

Gaskell, T. F., and Martin Morris, *World Climate: The Weather, the Environment and Man.* London: Thames & Hudson, 1979.

Gedzelman, Stanley David, *The Science and Wonders of the Atmosphere.* John Wiley & Sons, 1980.

Geikie, Sir Archibald, *The Founders of Geology.* Dover Publications, 1962.

Geikie, James, *The Great Ice Age and Its Relation to the Antiquity of Man.* D. Appleton, 1874.

Goudie, A. S., *Environmental Change.* Oxford University Press, 1977.

Gribbin, John, *Future Weather and the Greenhouse Effect.* Delacorte Press/Eleanor Friede, 1982.

Gribbin, John (ed.), *Climatic Change.* Cambridge University Press, 1978.

Gribbin, John, and Jeremy Cherfas, *The Monkey Puzzle: Reshaping the Evolutionary Tree.* Pantheon, 1982.

Hadingham, Evan, *Secrets of the Ice Age: The World of the Cave Artists.* Walker and Company, 1979.

Hambrey, M. J., and W. B. Harland (eds.), *Earth's Pre-Pleistocene Glacial Record.* Cambridge University Press, 1981.

Hamelin, Louis-Edmond, and Frank A. Cook, *Le Périglaciaire par l'Image: Illustrated Glossary of Periglacial Phenomena.* Quebec: Les Presses de L'Université Laval, 1967.

Herbert, A. P., *The Thames.* London: Weidenfeld and Nicolson, 1966.

Hoyle, Fred, *Ice: The Ultimate Human Catastrophe.* Continuum, 1981.

Imbrie, John, and Katherine Palmer Imbrie, *Ice Ages: Solving the Mystery.* Enslow Publishers, 1979.

Irons, James Campbell, *Autobiographical Sketch of James Croll, with Memoir of his Life and Work.* London: Edward Stanford, 1896.

Jelínek, J., *The Pictorial Encyclopedia of The Evolution of Man.* Hamlyn, 1975.

John, Brian S.:
 The Ice Age: Past and Present. London: Collins, 1977.

The World of Ice: The Natural History of the Frozen Regions. London: Orbis Publishing, 1979.

John, Brian S. (ed.), *The Winters of the World: Earth under the Ice Ages.* John Wiley & Sons, 1979.

Kahlke, Hans Dietrich, *Das Eiszeitalter.* Leipzig: Urania-Verlag, 1981.

Klein, Richard G.:
 Ice-Age Hunters of the Ukraine. University of Chicago Press, 1973.
 Man and Culture in the Late Pleistocene. Chandler Publishing, 1969.

Kurtén, Björn:
 The Age of Mammals. Columbia University Press, 1972.
 The Ice Age. Putnam, 1972.
 Pleistocene Mammals of Europe. Aldine, 1968.

Kurtén, Björn, and Elaine Anderson, *Pleistocene Mammals of North America.* Columbia University Press, 1980.

Lamb, H. H.:
 The Changing Climate. London: Methuen, 1966.
 Climate, History and the Modern World. Methuen, 1982.
 Climate: Past, Present and Future, Vol. 2. London: Methuen, 1977.

Latham, Robert, and William Mathews (eds.), *The Diary of Samuel Pepys,* Vol. 4, *1663.* University of California Press, 1971.

Leakey, Richard E., *The Making of Mankind.* E. P. Dutton, 1981.

Lewin, Roger, *Thread of Life.* W. W. Norton, 1982.

Lurie, Edward, *Louis Agassiz: A Life in Science.* University of Chicago Press, 1960.

McPherson, John G., *Footprints Frozen Continent.* London: Methuen, 1975.

Martin, P. S., and H. E. Wright Jr. (eds.), *Pleistocene Extinctions: The Search for a Cause.* Yale University Press, 1967.

Mather, Kirtley F., and Shirley L. Mason, *A Source Book in Geology.* McGraw-Hill, 1939.

Matsch, Charles L., *North America and the Great Ice Age.* McGraw-Hill, 1976.

Michael, Henry N., and Elizabeth K. Ralph (eds.), *Dating Techniques for the Archaeologist.* Massachusetts Institute of Technology, 1971.

Milankovitch, Milutin:
 Canon of Insolation and the Ice-Age Problem. Transl. by Israel Program for Scientific Translation. Belgrade: Royal Serbian Academy, 1941.
 Durch ferne Welten und Zeiten. Leipzig: Roehler & Umelang, 1936.

National Research Council, *Understanding Climatic Change: A Program for Action.* National Academy of Sciences, 1975.

Neale, John, and John Flenley (eds.), *The Quaternary in Britain.* Pergamon Press, 1981.

Péwé, Troy L. (ed.), *The Periglacial Environment: Past and Present.* Montreal: McGill-Queen's University Press, 1969.

Pfeiffer, John E., *The Emergence of Man.* Harper & Row, 1969.

Pielou, E. C., *Biogeography.* John Wiley & Sons, 1979.

Ponte, Lowell, *The Cooling.* Prentice-Hall, 1977.

Schultz, Gwen, *Ice Age Lost.* Anchor Press/Doubleday, 1974.

Silverberg, Robert, *Mammoths, Mastodons and Man.* McGraw-Hill, 1970.

Smith, A. G., et al., *Phanerozoic Paleocontinental World Maps.* Cambridge University Press, 1981.

Smithsonian Book of the Sun. W. W. Norton, 1981.

Solecki, Ralph S., *Shanidar: The First Flower People.* Alfred A. Knopf, 1971.

Sugden, David E., and Brian S. John, *Glaciers and Landscape: A Geomorphological Approach.* Edward Arnold, 1976.

Vereshchagin, N. K., *Zapiski Paleontologa Po Sledan Predkob.* Leningrad: Academy of Science, 1981.

Vereshchagin, N. K. (ed.), *Mlekopintaiushchie Vostochnoi Evrody B Antpogene.* Leningrad: Academy of Science, 1981.

Von Koenigswald, Wighart, and Joachim Hahn, *Jagdtiere und Jäger der Eiszeit.* Stuttgart: Konrad Theiss Verlag, 1981.

Warlow, Peter, *The Reversing Earth.* London: J. M. Dent & Sons, 1982.

Washburn, A. L., *Geocryology: A Survey of Periglacial Processes and Environments.* John Wiley & Sons, 1973.

Wenke, Robert J., *Patterns in Prehistory: Mankind's First Three Million Years.* Oxford University Press, 1980.

Wigley, T.M.L., M. J. Igram and G. Farmer (eds.), *Climate and History: Studies in Past Climates and Their Impact on Man.* Cambridge University Press, 1981.

Zeuner, Frederick E., *Dating the Past: An Introduction to Geochronology.* Hafner Publishing, 1970.

Periodicals

"Anti Matter." *Omni,* September 1982.

Beaty, Chester B., "The Causes of Glaciation." *American Scientist,* July-August 1978.

"Bolivian Lake Dated Back to Last Ice Age." *On Campus* (Ohio State University), October 7, 1982.

Broecker, Wallace S., "Climatic Change: Are We on the Brink of a Pronounced Global Warming?" *Science,* August 8, 1975.

Broecker, W. S., et al., "Milankovitch Hypothesis Supported by Precise Dating of Coral Reefs and Deep-Sea Sediments." *Science,* January 19, 1968.

Brownlee, Donald E., "Cosmic Dust." *Natural History,* April 1981.

Bryson, Reid A., and John E. Ross, "Climatic Variation and Implications for World Food Production." *World Development,* May-July 1977.

CLIMAP project members, "The Surface of the Ice-Age Earth." *Science,* March 19, 1976.

Cragin, Jim, "Tales the Ice Can Tell." *Mosaic,* September/October 1978.

Croll, James, "On the Excentricity of the Earth's Orbit, and Its Physical Relations to the Glacial Epoch." *London, Edinburgh, and Dublin Philosophical Magazine and Journal of Science.* January-June 1867.

"Cultural Evolution." *Mosaic,* March/April 1979.

Dansgaard, W., "Ice Core Studies: Dating the Past to Find the Future." *Nature,* April 2, 1981.

Dansgaard, W., et al., "A New Greenland Deep Ice Core." *Science,* December 24, 1982.

Dansgaard, Willi, and Jean-Claude Duplessy, "The Eemian Interglacial and Its Termination." *Boreas,* February 1981.

Diamond, Jared M., "Man the Exterminator." *News and Views,* August 26, 1982.

Domico, Terry, "Summer on the Glacier." *Alaskafest,* July 1982.

Eddy, John A., "The Case of the Missing Sunspots." *Scientific American,* May 1977.

Emiliani, Cesare:
 "Ice Sheets and Ice Melts." *Natural History,* November 1980.
 "Pleistocene Temperatures." *Journal of Geology,* November 1955.

Evans, J. V., "The Sun's Influence on the Earth's Atmosphere and Interplanetary Space." *Science,* April 30, 1982.

"Extinctions and Ice Ages: Are Comets to Blame?" *New Scientist,* June 10, 1982.

Fairbridge, Rhodes W.:
 "Early Paleozoic South Pole in Northwest Africa." *Geological Society of America Bulletin,* January 1969.
 "Upper Ordovician Glaciation in Northwest Africa? Reply." *Geological Society of America Bulletin,* January 1971.

Ferguson, C. W., "Bristlecone Pine: Science and Esthetics." *Science,* February 23, 1968.

Fodor, R. V., "Frozen Earth: Explaining the Ice Ages." *Weatherwise,* June 1982.

Gates, W. Lawrence, "Modeling the Ice-Age Climate." *Science*, March 1976.

Goldthwait, Richard P., "The Growth of Glacial Geology and Glaciology: Opening Remarks." *Geoscience Canada*, March 1982.

Gribbin, John:
"Stand By for Bad Winters." *New Scientist*, October 28, 1982.
"Sun and Weather: The Stratospheric Link." *New Scientist*, September 10, 1981.

Guthrie, Russell D., "Re-creating a Vanished World." *National Geographic*, March 1972.

Hammer, C. U., et al., "Dating of Greenland Ice Cores by Flow Models, Isotopes, Volcanic Debris, and Continental Dust." *Journal of Geology*, Vol. 20, No. 82.

Hansen, J., et al., "Climate Impact of Increasing Atmospheric Carbon Dioxide." *Science*, August 28, 1981.

Hays, J. D., et al., "Variations in the Earth's Orbit: Pacemaker of the Ice Ages." *Science*, December 10, 1976.

Hoinkes, Herfried C., "Surges of the Vernagtferner in the Ötztal Alps since 1599." *Canadian Journal of Earth Sciences*, August 1969.

Hudson, J. Harold, et al., "Sclerochronology: A tool for interpreting past environments." *Geology*, Vol. 4.

Imbrie, John, "Astronomical Theory of the Pleistocene Ice Ages: A Brief Historical Review." *Icarus*, May/June 1982.

Kerr, Richard A.:
"El Chichón Forebodes Climate Change." *Science*, September 10, 1982.
"Milankovitch Climate Cycles: Old and Unsteady." *Science*, September 4, 1981.

Klein, Jeffrey, et al., "Calibration of Radiocarbon Dates." *Radiocarbon*, Vol. 24, No. 2.

Krantz, Grover S., "Human Activities and Megafaunal Extinctions." *American Scientist*, March-April 1970.

Kukla, George J., "Around the Ice Age World." *Natural History*, April 1976.

Kukla, George J., and Jeffrey A. Brown, "Impact of Snow on Surface Brightness." *EOS*, July 20, 1982.

Kurtén, Björn, "The Cave Bear." *Scientific American*, March 1972.

Kutzbach, John E., "Monsoon Climate of the Early Holocene: Climate Experiment with the Earth's Orbital Parameters for 9,000 Years Ago." *Science*, October 2, 1981.

Livingstone, D. A., "Late Quaternary Climatic Change in Africa." *Annual Review of Ecology and Systematics*, Vol. 6, 1975.

McCauley, J. F., et al., "Subsurface Valleys and Geoarcheology of the Eastern Sahara Revealed by Shuttle Radar." *Science*, December 3, 1982.

Maran, Stephen P., "The Inconstant Sun." *Natural History*, April 1982.

Marsh, Peter, and Mike Winney, "London Rolls Back the Tide of the Thames." *New Scientist*, November 5, 1981.

Marshack, Alexander, "Exploring the Mind of Ice Age Man." *National Geographic*, January 1975.

Marshall, Larry G., et al., "Mammalian Evolution and the Great American Interchange." *Science*, March 12, 1982.

Martin, Paul S., "Pleistocene Overkill." *Natural History*, December 1967.

Matthews, Samuel W., "What's Happening to Our Climate?" *National Geographic*, November 1976.

Mosley-Thompson, E., and L. G. Thompson, "Nine Centuries of Microparticle Deposition at the South Pole." *Quaternary Research* 17, 1982.

Neftel, A., et al., "Ice Core Sample Measurements Give Atmospheric CO_2 Content during the Past 40,000 yr." *Nature*, January 21, 1982.

"The Next Ice Age May Be Closer Than You Think." *New Scientist*, July 22, 1982.

Raisbeck, G. M., et al., "Cosmogenic [10]Be Concentrations in Antarctic Ice during the Past 30,000 years." *Nature*, August 27, 1981.

Reader, John, "The Beckoning Snows of Kilimanjaro, Africa's 'Mountain Greatness.'" *Smithsonian*, August 1982.

Reiners, William A., et al., "Plant Diversity in a Chronosequence at Glacier Bay, Alaska." *Ecology*, Winter 1971.

Rensberger, Boyce:
"The Emergence of *Homo Sapiens*." *Mosaic*, November/December 1980.
"Facing the Past." *Science 81*, October 1981.

Risbo, T., et al., "Supernovae and Nitrate in the Greenland Ice Sheet." *Nature*, December 17, 1981.

Ruddiman, William F., "North Atlantic Ice-Rafting: A Major Change at 75,000 Years before the Present." *Science*, June 10, 1977.

Ruddiman, William F., and Andrew McIntyre:
"Oceanic Mechanisms for Amplication of the 23,000-Year Ice-Volume Cycle." *Science*, May 8, 1981.
"Warmth of the Subpolar North Atlantic Ocean during Northern Hemisphere Ice-Sheet Growth." *Science*, April 13, 1979.

Schmid, Rudolf, and Marvin J. Schmid, "Living Links with the Past." *Natural History*, March 1975.

Schultz, P., and D. Gault, "Cosmic Dust & Impact events." *Geotimes*, June 1982.

Scotese, Christopher R., et al., "Paleozoic Base Maps." *The Journal of Geology*, May 1979.

Shinn, Eugene A., "Time Capsules in the Sea." *Sea Frontiers*, November-December 1981.

Sigurdsson, Haraldur, "Volcanic Pollution and Climate: The 1783 Laki Eruption." *EOS*. August 10, 1982.

"South Pole Reaches the Sahara." *Science*, May 15, 1970.

Stewart, John Massey:
"A Baby That Died 40,000 Years Ago Reveals a Story." *Smithsonian*, September 1979.
"Frozen Mammoths from Siberia Bring the Ice Ages to Vivid Life." *Smithsonian*, December 1977.

Sullivan, Walter, "Ancient Ice Yielding Secrets of Climate." *The New York Times*, August 8, 1981.

Thorarinsson, Sigurdur, "Glacier Surges in Iceland, with Special Reference to the Surges of Brúarjökull." *Canadian Journal of Earth Sciences*, August 1969.

Weertman, Johannes, "Milankovitch Solar Radiation Variations and Ice Age Ice Sheet Sizes." *Nature*, May 6, 1976.

Weintraub, Boris, "Fire and Ash, Darkness at Noon." *National Geographic*, November 1982.

Wetmore, Alexander, "Re-creating Madagascar's Giant Extinct Bird." *National Geographic*, October 1967.

Wholey, Jane, "Saving London from an Impending Threat of flood." *Smithsonian*, August 1982.

Wilson, A. T., "Origin of Ice Ages: An Ice Shelf Theory for Pleistocene Glaciation." *Nature*, January 11, 1964.

Woillard, Geneviève, "Abrupt End of the Last Interglacial S.S. in North-east France." *Nature*, October 18, 1979.

Wolkomir, Richard, "Waging War against the Cold Is the Job of a Unique Army Lab." *Smithsonian*, February 1981.

Woodward, H. B., "Dr. Buckland and the Glacial Theory." *Midland Naturalist*, Vol. 6, 1883.

Wright, G. Frederick, "Agassiz and the Ice Age." *American Naturalist*, March 1898.

Zimmerman, P. R., et al., "Termites: A Potentially Large Source of Atmospheric Methane, Carbon Dioxide, and Molecular Hydrogen. *Science*, November 5, 1982.

Zwally, H. Jay, and Per Gloersen, "Passive Microwave Images of the Polar Regions and Research Applications." *Polar Record*, Vol. 18, No. 166.

Other Publications

Bentley, Charles, "Carbon Dioxide Effects: Research and Assessment Program—Environmental and Societal Consequences of a Possible CO_2 Induced Climate Change," Vol. 2, Part 1. A project conducted by the American Association for the Advancement of Science for the U.S. Department of Energy, April 1982.

Birchfield, G. E., and Johannes Weertman, "A Model Study of the Role of Variable Ice Albedo in the Climate Response of the Earth to Orbital Variations." November 16, 1981.

CLIMAP project members:
"Glacial North Atlantic 18,000 Years Ago: A CLIMAP Reconstruction." Geological Society of America, Memoir 145, 1976.
"Seasonal Reconstructions of the Earth's Surface at the Last Glacial Maximum." Geological Society of America Map and Chart Series, MC-36, 1981.

Dansgaard, W., et al., "Dating and Climatic Interpretation of the Dye 3 Deep Ice Core." University of Copenhagen, no date.

Fairbridge, Rhodes W., "Glacial Grooves and Periglacial Features in the Saharan Ordovician." *Glacial Geomorphology*, State University of New York, September 1974.

"Glacial Epoch." *McGraw-Hill Encyclopedia of Science & Technology*, 5th ed., 1982.

Goldthwait, R. P., et al., *Soil Development and Ecological Succession in a Deglaciated Area of Muir Inlet, Southeast Alaska*. Institute of Polar Studies, Report No. 20, June 1966.

Greater London Council:
"The GLC Thames Flood Barrier." January 1981.
"How the Thames Barrier Will Work." September 1981.

"The Great Ice Age." U.S. Department of the Interior/Geological Survey pamphlet, 1978.

Herron, Michael M., and Chester C. Langway Jr., "Chloride, Nitrate, and Sulfate in the Dye 3 and Camp Century, Greenland Ice Cores." AGU/GISP Symposium, June 2, 1982, Philadelphia.

Hudson, J. Harold, "Response of *Montastraea Annularis* to Environmental Change in the Florida Keys." 4th Annual Coral Reef Symposium, 1981.

Hudson, J. H., et al., "Effects of Offshore Oil Drilling on Philippine Reef Corals." Preprint: *Bulletin of Marine Science*, Vol. 32, No. 4, October 1982.

Keck, W. G., and W. R. Hassibe, "The Great Salt Lake." U.S. Department of the Interior/Geological Survey pamphlet, no date.

Lawrence, Donald B., "Primary Versus Secondary Succession at Glacier Bay National Monument, Southeastern Alaska." *Proceedings of the First Conference on Scientific Research in the National Parks, New Orleans, Louisiana, November 9-12, 1976.*

North, F. J., "Centenary of the Glacial Theory." *Proceedings of the Geologists' Association*. London: Edward Stanford, March 26, 1943.

Péwé, Troy L.:
"Permafrost: Challenge of the Arctic." *1976 Yearbook of Science and the Future*. Encyclopedia Britannica, 1975.

Pratt, S. W., and J. M. Holloway, "Thames Barrier Project: Provision of Services." Paper presented to the Building Construction Forum of the Greater London Council, February 28, 1978.

Proceedings of the Geological Society of London, Vol. 3, Part 2, 1840-1841, No. 72.

Svoboda, Josef, and Bill Freedman, "Ecology of a High Arctic Lowland Oasis: Alexandra Fiord (78°53'N. 75°55'W.), Ellesmere Island, N.W.T., Canada." *Second Annual Report of the Alexandra Fiord Lowland Ecosystem Study*. Departments of Botany, University of Toronto and Dalhousie University, November 1981.

Thomas, Robert H., and H. Jay Zwally, "Space Surveillance of Changes in Polar Ice." July 14, 1982.

Thompson, L. G., and E. Mosley-Thompson, "Temporal Variability of Microparticle Properties in Polar Ice Sheets." *Journal of Volcanology and Geothermal Research*. Amsterdam: Elsevier, 1981.

Webb, Peter-Noel, "Review of Late Cretaceous-Cenozoic Geology of the Ross Sector, Antarctica." Fourth International Symposium on Antarctic Earth Sciences, Adelaide, Australia, August 1982.

Webb, S. David, "Underwater Paleontology of Florida's Rivers." National Geographic Society Research Reports, 1968.

Zwally, H. J., et al.:
"Ice-Sheet Dynamics by Satellite Laser Altimetry." National Aeronautics and Space Administration, Technical Memorandum 82128, May 1981.

"Variability of Antarctic Sea Ice and CO$_2$ Change." September 10, 1982.

Exhibits
"Ice Age Mammals and the Emergence of Man." An exhibit at the National Museum of Natural History, Smithsonian Institution, Washington, D.C.

PICTURE CREDITS

INDEX

Orbital path of earth, 100, 101, *102,* 158, 159; eccentricity and cycle, *102,* 104, *diagram* 106, 108-109, 135, 137

Orcutt, W. W., 60-61

Ordovician period, 141, 142

Oxygen-isotope analysis, 131, 132, 135, 137, 151-152, 153-154, 155, 158, 161, 163

P

Pacific Ocean: Ice Age temperatures of, 29; sediment analysis, 135, 137

Paleomagnetism, 139, 141

Papua New Guinea, sea-level changes, *134*

Patagonia, 24

Patterned-ground polygons, 30, *32*

Peccary, 57, 67, 69

Peirce, Benjamin, *84*

Penck, Albrecht, 97, 109, 119, 120

Pepys, Samuel, 164

Permafrost, 19, 30, 70, 71; landforms, 30, *31-34;* last ice age, 30

Perraudin, Jean Pierre, 85

Peru, tar-pit fossils, 61

Philippines, 28; coral reef, *133*

Pilgrim, Ludwig, 108

Pines, 30, 35, 167; bristlecone, *118*

Pingos, 30, *34*

Planetary synod, 163

Plankton, 151, 152

Plant life: Ice Age, 30, 35; Ice Age vs. interglacial, 166-167; revegetation, 41, *42-51;* tundra, 30, 58

Pleistocene ice epoch, 18, 19, 20, *graph* 21, 22, 57, 69, 149, 156, 169; beginning of, 135; chronology research, 109, 119, 120, 129, 135, 137-139; duration of, 97; megafauna, 62, *64-66,* 67, 69-70, *71, 72-75, 76-81;* migrations, 62, 67

Poland, in Ice Age, 24, 30

Polar regions, 109; ice-sheet drilling, *110-117;* reflection of solar energy by, 100

Potassium-argon dating, 134

Precipitation: Ice Age, 29-30, 35; postglacial, 158-159. *See also* Snow

Pressure melting point, 121, 122

R

Racial senility concept, 70, 72

Radioactive dating, 110, 119-120, 134, 137, 154. *See also* Carbon-14 dating

Rainfall: Ice Age, 29-30, 35; postglacial, 158-159; sunspot activity and, 124

Reindeer, 35, 38, 61, 74-75

Ren Zhenqiu, 163

Revegetation, 41, *42-51*

Rhinoceros, 56, 67; woolly, 57, 62, *64-65,* 67, 69, 74

Riss ice age, 97

Rock, striated, 83, 87, *88,* 89, 141, 142, *144*

Rockies: Canadian, 24; U.S., 159

Ross Ice Shelf, *110-111, 112*

Royal Society (London), 105

Rubin, Meyer, 120

Ruddiman, William F., 151, 152, 155

Russia: Cro-Magnons, 35; permafrost, 30; stampede hunting, 17-18, 73; tar-pit fossils, 61

Ruwenzori mountain range, 156

S

Saber-toothed cat, 56, 60-61, *66,* 67, 69, 76-77

Sahara, 30; cave drawings, 158; Ordovician glacial evidence in, 141, *142-147;* postglacial climate changes, 158, 159

St. Helens, Mount, 125

St. John's Bay, Newfoundland, *88*

Salt flats, *96*

Scandinavia, 167; butterfly-relict habitat, *map* 72; Lapps of, 75; marine fossils in mountains, 90; myths, 83; permafrost, 30

Scandinavian Ice Sheet, 24, 35, 95

Schaefer, Ingo, 119

Schimper, Karl, 84, 87

Schott, Wolfgang, 130

Scotland, Ice Age evidence in, 87, 89, 90

Sea level, changes in, 90, 132, *134;* drop in Ice Age, 20, 24, 28, 29, 90; effect on land animals, 67; postglacial rise, 90, 92, 94, 132, 158, 164; potential rise in warm-up, 19

Seasons: axial tilt and, 100-101, *diagram* 107, 158, 159; orbital eccentricity and, *102,* 104, 158, 159; present trends, 159; wobble and, *diagram* 107

Sediment, glacial (till), 29, 90, 93, *146;* in lakes, varves, 95

Shackleton, Nicholas, 132, 135, 137

Siberia: fossil finds, 58, *71;* in Ice Age, 24, 30; mammoth, 54, 55-56, 70, *71;* megafauna, 55

Sloths, 69, 70. *See also* Ground sloth

Smilodon, *66,* 76-77

Smithsonian Institution, 76, 108

Snow: accumulation, 155, 159, 163; accumulation layers, *148,* 153-154; Ice Age scarcity, 155; of Maunder minimum, 163; oxygen-isotope analysis, 152-155; reflection of solar energy by, 97, 151, 167

Snowblitz theory of ice ages, 122, 129

Society of American Archeology, 75

Soergel, Wolfgang, *73*

Solar constant, 108, 161, 163

Solar Maximum Mission satellite, 161, 163

Solar radiation, 97, 100, 121, 122, 150, 163; fluctuation in amount received, 102, 104, 106, *diagram* 107, 108-109, 137-138, 152, 158, 159; meteorite-dust shield, 139; Milankovitch curve, *diagram* 106, 109, 120, 129, 132; reflected by snow, 97, 151, 167; volcanic-dust shield, 123-124, 155, 161

Sonoran desert, 35

South Africa, pre-Pleistocene ice-age traces in, 139

South America: extinction of species in, 57; fossil finds, 57, 61; giant mammals, 62; glaciers of Ice Age, 24; migratory exchange of fauna with North America, 67; pre-Pleistocene ice-age traces in, 139, 142

South Devon, cave fossils, 62

Southern Hemisphere, 121, 138; ice ages of, 104; in last ice age, 24; pre-Pleistocene ice-age traces in, 139; seasons, 100-101

South Pole, 101; Ordovician land area, 141, 142

Spruce, 30, 35, 41, *46-49,* 166-167

Steppe, 30, 35, 70

Stratosphere: meteorite-impact dust in, 138, 139; volcanic-emission profile, *129*

Striations, glacial, 83, 87, *88,* 89, *144*

Strommel, Henry and Elizabeth, 123

Sub-Arctic region, permafrost, 30

Subtropics, postglacial climate changes, 158-159

Sulfur dioxide gas, 125

Summer, 100-101, 107, 158; diminution of heat, temperate zone, and ice age, 109; Milankovitch curve, 109; present trend, 159

Sunspots, 124, 129, 139, 163

Surge, glacial, 122, 129

Swiss Society of Natural Sciences, 83, 85, 86, 167

Switzerland: Alps, 85, 97; Jura, 83, 85

T

Tambora, Mount, 123-124

Tapir, 57, 62, 67, 74, 76

Tar-pit fossils, *60-61*

Tasmania, 24

Temperate zone: forests, 166-167; heat diminution and glaciation, 109

Temperatures, 161; climatic optimum, 158; coldest, last ice age, 29; present-day global average, 20; present interglacial fluctuations, 158-159. *See also* Ocean temperatures

Thames River, storm-surge gates, *164-165*

Thompson, Lonnie, 155

Tien Shan, in Ice Age, 24

Till, glacial, 90, 95, 97; in Sahara, *146*

Timocharis, 101

Trees: Ice Age, 30, 35; interglacial vs. glacial, 166-167; in revegetation, 41, *44-51*

Trimmer, Joshua, 90

Tropics, 97, 100; during Ice Age, 29; postglacial climate changes, 158, 159; pre-Pleistocene glacial trees found in, 139, 141

Tundra, 30, *58,* 70; fossil depository, 58, *59*

Tunisia, 30

U

Uganda, 24, 156

Uniformitarianism, law of, 85, 95

United States: Cordilleran Ice Sheet, 24, 28; Ice Age temperature, 29; Laurentide Ice Sheet, 24, 35; retreat of ice, 35. *See also* North America; *individual states*

U.S. Geological Survey, 90, 160

U.S. National Aeronautics and Space Administration (NASA), 159, 161

University of Arizona, 73

University of Belgrade, 105

University of Bern, 158

University of Chicago, 119, 131

University of Minnesota, 75

University of New Mexico, 70

University of Toronto, 155

University of Wisconsin, 158

Uplift, postglacial, 6, *10-11,* 90, *164*

Ural Mountains, 24

Urey, Harold, 131

Ussher, Archbishop James, 85

Utah, 30, 90; salt flats, 96

V

Valleys, glacial, 28

Varves, 95, 97

Venetz, Ignatz, 83, 85, 86

Volcanic eruptions, 123-124, *125,* 129, 139, 155, 161; atmospheric effects studied, *126-129;* debris in ice, *116,* 152, 155

W

Wallace, Alfred Russel, 55, 56-57, 72, 75

Washington State University, 74

Water, amount locked in ice: Ice Age, 24; today, 19

Water vapor, atmospheric: and diamond dust, 138-139; as radiation trap, 122-123

Weertman, Johannes, 149

Wegener, Alfred, 109, 121, 139

West Antarctic marine ice sheet, 160

Whittlesey, Charles, 90

Wilson, Alex T., 121-122

Winds, Ice Age, 29, 151, 155

Winter, 100-101, 107, 109; at aphelion vs. perihelion, *102,* 104, 159; present trends, 159

Wisconsinan ice age, 97

Wobble, 101, 104, 106, *diagram* 107, 134. *See also* Axial precession

Woillard, Geneviève, 166-167

Wolf, dire, 60-61, 69, 79

Wolverine, 62

Woodroffe, Alastair, 122

Woolly mammoth, 56, 57, 61, *64-65,* 67, 70, *71, 81*

Woolly rhinoceros, 57, 62, *64-65,* 67, 69, 74

Würm ice age, 97

X

X-ray emission, *166*

Y

Yale University, 120

northwest Greenland. Released from the burden of the ice as it receded, sections of the coast have risen some 390 feet during the past 12,000 years.

An advancing plain of ice flows into the sea from the Greenland Ice Sheet as waves and tides chisel icebergs from its glacial snout.

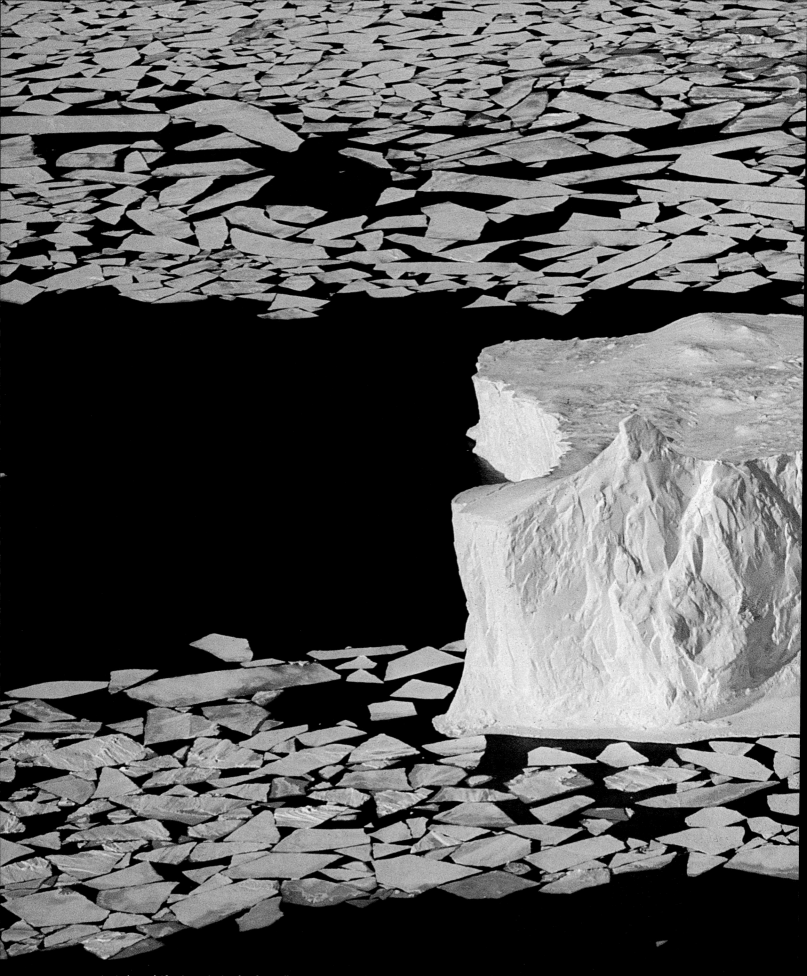

An iceberg drifts through shards of ice off the west coast of Greenland. Icebergs and other bodies of floating ice are currently confined largely to polar

seas, but at the height of the last ice age they choked half of the world's oceans.

Chapter 1

A FRIGID AGE OF CHANGE

For as long as anyone could remember, the band of hunters—along with women and children—had spent a part of each year near the sheer limestone cliffs watching for game animals to appear. The place where they waited lay on a peninsula jutting into what would be known thousands of years later as the Black Sea. To the north, a cliff plunged some 50 feet into a narrow valley; to the east and west, deep gullies had been cut into the rock by intermittent streams. It was an ideal natural location for the enterprise that the hunters had in mind.

One day, some of the men who had gone out in search of prey came rushing back to the camp to report that a herd of grazing donkeys was heading in their direction. Quickly, the men took up their prearranged posts. Some of them, gripping flint-headed spears, rocks or stout wooden clubs, clambered down the nearest gully and took up positions at the foot of the steep cliff. Others, carrying long sticks in each hand, trotted away from the precipice and concealed themselves in the clumps of trees that punctuated the grassy plateau. Shivering from cold and excitement, they waited.

Several hours later, they saw their quarry, several dozen thin-legged donkeys browsing in the grasses that, even in winter, sustained a variety of grazing animals. Staying close together while they nibbled at the grass, the donkeys edged ever closer to the cliffs and gullies. Suddenly, one of the men—the leader of the hunt—leaped from his hiding place and began to run toward the herd, beating his sticks together and whooping as he went. The other hunters followed immediately, spreading out across the plateau between the two gullies and running toward the steep cliff. The donkeys, startled by the din of shrill cries and clattering sticks, milled about in panic, then fled from the rapidly advancing line of men. Some of the stampeding animals tried to run to the left or right but turned back when they neared the precipitous gullies. Soon, the whole herd was thundering straight ahead, toward the edge of the cliff.

As the animals neared the precipice, the screeching hunters on the flanks ran faster, turning the onrushing line of men into a tightening half circle. The donkeys, their piercing brays cutting through the cold winter air, were getting closer and closer to the cliff's edge now; a few of them, seeing the danger ahead, wheeled about and dashed through the cordon of pursuers. But for most of the herd, there was no escape; they galloped headlong off the cliff or were shoved over the edge by the panting men behind them.

Most of the animals died when they hit the rocky floor of the valley below. Those that survived, their frail legs splintered by the fall, were

Dimly visible on the wall of Gargas Cave in the French Pyrenees, outlined hands of Ice Age artists signal across a span of at least 12,000 years. Such hand images, created by the simple tactic of blowing powdered pigments over fingers placed against rock, are frequently found in the Cro-Magnon cave art of Europe.

quickly clubbed or speared to death by the men who had been waiting at the foot of the cliff. When the killing was finished, the men hauled the donkey carcasses up the gulley to the camp, where they butchered them with flint knives.

This large-scale hunt—its description based on fossil evidence that was discovered in 1953—took place on the Crimean peninsula some 30,000 years ago, when the earth was enduring extensive glaciation. The hunters lived far from the massive ice sheets that had intruded deeply into the Northern Hemisphere. But their world—its climate, and therefore its plant and animal life—was shaped in countless ways by the great glaciers that pressed toward the lower latitudes, withdrawing from time to time only to move outward once more.

By the time those Crimean hunters staged their highly organized donkey hunts, humanity had been in existence for several million years—although for most of that period creatures who bore the name *Homo* were decidedly less intelligent and resourceful. Now human society was developing at a rapidly accelerating pace. Humans had evolved to the point of full modernity: They lived by their wits and accumulated learning to a degree that was far beyond the powers of even their recent ancestors.

Civilization is considered by some to be a product of the encroachments of ice. Language, cooperation, the ability to plan ahead—the very skills that enabled ancient hunters to drive a donkey herd toward a deadly precipice—may have evolved partly in response to the global changes in climate that accompanied periods of glacial advance and retreat. One anthropologist, Charles K. Brain of South Africa, has maintained that had it not been for the challenges posed by such environmental changes, humanity might still be locked in an early stage of evolutionary development, living a simple life in the tropics.

The cold period that saw the rise of modern humans was hardly a unique and freakish event. Certain ancient rock formations bear the imprint of massive glacial action that is estimated to have occurred more than two billion years ago; within the last billion years, at least six other ice ages have taken place—apparently spaced at intervals of about 150 million years and lasting for as long as 50 million years. The very term "ice age" can be confusing, since it is generally used to refer to a period of sustained glaciation lasting a few tens of thousands of years. When scientists use the term, however, they frequently mean a period of global cooling that can last for millions of years and consists of a series of glacial advances and retreats, called glacials and interglacials. As a practical matter, it seems reasonable to call an event of such long duration an "ice era" or an "ice epoch" and to reserve the term "ice age" for the relatively brief periods of most severe cooling.

The most recent ice era began about 65 million years ago. Its effects were mild at first. About 55 million years ago, small glaciers began to form in Antarctica. Growing for a time, then shrinking and growing again, they gradually expanded and coalesced into a dome-shaped ice sheet that by 20 million years ago had spread out so that it covered the whole continent. Twelve million years ago, glaciers appeared on the mountains of Alaska; by three million years ago, Greenland was covered by an ice sheet. Since the beginning of the geological period known as the Pleistocene epoch, about two and a half million years ago, the ice sheets of North America and Europe have advanced far into the middle latitudes at least four times, and perhaps as many as 10 times. Each of these glacial incursions could be considered a full-scale ice age. In addition, there have been at least 10 other occasions when the ice sheets spread far beyond their present extent.

Even now, in an interglacial period, the ice era continues; glaciers still hold a considerable part of the earth in frigid thrall. Almost six million cubic miles of ice burden the continent of Antarctica, and more than half a million cubic miles blanket Greenland and other parts of the north polar regions. Altogether, about 10 per cent of the earth's land area is covered by ice sheets and glaciers; another 14 per cent is affected by ice in the ground—permafrost. If all the ice on earth were to melt, so much water would be released that the global sea level would rise by more than 200 feet, flooding many of the world's major cities.

Not much is known about the ice ages that occurred before the Pleistocene ice epoch. There can be no doubt, however, that the course of evolution was profoundly affected by periods of severe cooling. The rise of the genus Homo is a case in point.

Perhaps as long as 10 million years ago, apelike human forebears lived and foraged for food in the lush tropical forests of Africa and Asia. But the warm and comfortable world of these creatures was slowly cooling as implacable sheets of ice spread out from the higher latitudes. As the global climate chilled, the hospitable forests began retreating toward the Equator. Their place was taken by open woodlands and vast grasslands. For a time, the bands of apelike animals remained in the shrinking forests, competing ever more intensely among themselves and with other forest dwellers for dwindling food supplies. Then, perhaps four or five million years ago, the first protohumans, or hominids, appeared in Africa and began to abandon the thick forests that had sheltered and nurtured their ancestors for millions of years. In gradually increasing numbers, they took to the open woods and then to the regions of grass.

At first, these newcomers must have seemed frail and ill equipped to survive in their new realm. They were not as fleet of foot as the horses, zebras and other grazing animals that ranged the grasslands, and they had neither sharp teeth nor claws to defend themselves against such predators as leopards and saber-toothed cats. But they carried within them the evolutionary seeds that would in time permit them to master their treeless environment—and many other environments besides.

During their long years as dwellers in tangled woodlands, the hominids had developed excellent color vision and depth perception that far exceeded the capacities of any of the grassland dwellers. They had also evolved hands and fingers that were adept at gripping tree branches and would soon prove well suited to the fashioning of tools and weapons. Equally important, they were evolving a permanently upright posture, which would free their hands for the use of those tools and weapons.

When they first ventured onto the grasslands, the protohumans may have survived partly by scavenging—eating meat that they stripped from the carcasses of animals brought down by four-footed predators. But they eventually made fresh kills of their own and, as an aid in bagging their quarry, began to grasp and use sticks and stones. They also discovered that

The Complex Rhythms of Cold

Human beings have never experienced the earth's normal climate. For most of its 4.6-billion-year existence, the planet has been inhospitably hot or dry and utterly devoid of glacial ice. Only seven times have major ice eras, averaging roughly 50 million years in length, introduced relatively cooler temperatures; humankind arose during the most recent of those periods.

The chart at right, a visual representation of the key terms and chronology of the science of ice ages, shows the known ice eras in blue on the top line. Each ice era encompassed several ice epochs—periods of still lower average temperatures. During the 65 million years of the current ice era (second line), six ice epochs have occurred.

When the Pleistocene ice epoch began about 2.4 million years ago (third line), advancing ice sheets marked the onset of one of the coldest climatic inter-

ludes experienced on earth. These periods of especially vigorous glacial advance are known as the ice ages.

The most recent ice age (fourth line), preceded and followed by warmer times —interglacials—began about 120,000 years ago. It reached a bitter extreme some 50,000 years later, slowly moderated, then brought severe cold again about 18,000 years ago. In the last 10,000 years—the Holocene interglacial—three sustained cold spells (fifth line) sent temperatures below the current global average of 59° F. One of them, a time of crop failure and famine called the Little Ice Age, ran from the 15th to the 19th Century (sixth line). The 100-year warming trend that ended the Little Ice Age extended through the early 1960s but, as noted on the last line of the chart, it was followed by yet another drop in average temperature, which has lasted up to the present.

4.6 BILLION YEARS: SEVEN ICE ERAS

At the peak of the last ice age, 18,000 years ago, glaciers sheathed large portions of the globe, shown in blue on the map below. With 17 million cubic miles of the world's water transformed into ice, sea level dropped 400 feet; the continents assumed different shapes (green) and land bridges emerged, notably the link between North America and Asia.

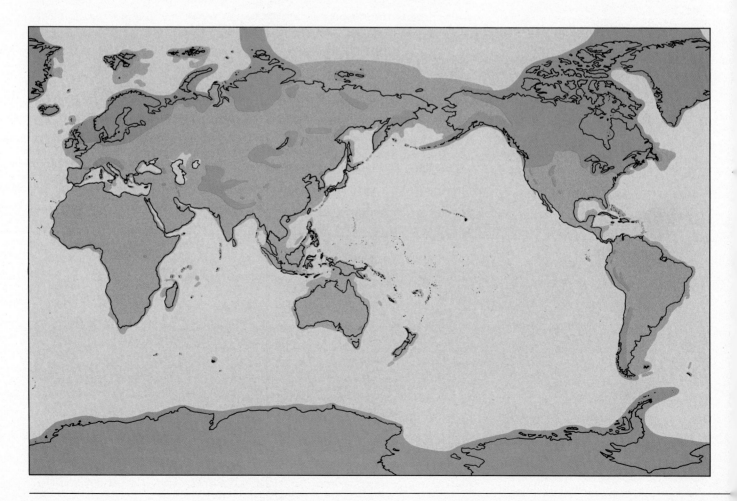

A series of time charts, each one embracing a fraction of the one before, depicts the swings between cold *(blue)* and warmth *(green)* that have characterized the climate of the earth for billions of years.

65 MILLION YEARS: AN ICE ERA

2.4 MILLION YEARS: AN ICE EPOCH

125,000 YEARS: AN ICE-AGE CYCLE

10,000 YEARS: AN INTERGLACIAL

1,000 YEARS

100 YEARS: 1880-1980

their hunts would be more successful if they cooperated with one another; cooperation required some form of communication, and it is probable that a rudimentary language developed as a response to the needs of group hunting and living.

By the middle of the Pleistocene epoch, about one million years ago, the protohumans had evolved into *Homo erectus*—fully upright man—with a brain much larger than that of any forebear, although only about two thirds of the modern brain size. In this form, human beings would spread out from Africa, eventually populating much of Europe and Asia. It is likely that the Pleistocene chill kept them from ranging much farther north than the latitudes of southern France or the Black Sea until, some 800,000 years ago, they learned to use and control fire. This precious gift would in time enable them to remain at high latitudes even when the world was plunged into ice-age conditions. All the while, *Homo erectus* was gaining the ability to plan and to manipulate the environment. About 300,000 years ago, humans reached the evolutionary level that scientists honor with the species name *Homo sapiens*—thinking man.

Some 100,000 years ago a variety of *Homo sapiens* known as Neanderthal—named for the German river valley in which the fossil remains were first found in 1856—strode onto the evolutionary stage. Little can be deduced from the scanty fossil record of the earliest generations, but by 70,000 years ago Neanderthals living in Europe were physically well adapted to periods of extreme cold. Their short and stocky build conserved body heat far more effectively than a tall and lanky frame would have done, and their large nose was suited to the task of warming the air they breathed. With their ability to control fire, to fashion tools of stone and to make rudimentary clothing out of animal skins, the Neanderthals were well prepared to contend with the great ice age that began shortly after they first appeared on the earth, a glaciation so momentous that it is formally known as the Ice Age.

About 125,000 years ago, temperatures throughout the Northern Hemisphere began a gradual decline. During the next 5,000 years, the volume of the ice sheets increased by some 1,200 cubic miles per year. Under the enormous weight of the new ice, the glacial masses began to deform, oozing inexorably outward. Their advance was not constant; on at least two occasions, during short-lived warmings of the world climate, they retreated. But the long-term climatic trend was ultimately downward, and by 65,000 years ago the ice sheets of the Northern Hemisphere covered most of Canada and Scandinavia.

Far from the towering sheets of advancing ice, winters became longer and colder, while the summers were cool and dry. Over time, many formerly forested regions of Europe were transformed into patchy woodlands, open grasslands and tundra. It was hardly an inviting winter habitat for human beings; during past ice ages, those who had penetrated northward into Europe had usually withdrawn from such inhospitable surroundings. But the Neanderthals were able to stand their ground, and this hardy breed flourished in the bitter cold of Western European winters for thousands of years, taking refuge in caves and rude shelters of animal bones and skins and subsisting on a variety of plant and animal foods.

These prototypical ice people were hardly the stooped, shuffling brutes of popular imagination *(pages 24-27)*. Indeed, their brains were slightly larger than those of modern humans, and they showed remarkable resourcefulness in dealing with a forbidding environment. Neanderthal hunters worked together to bag not only small animals, but also such big game as bison, elk and reindeer; and they moved their encampments with the shifting seasons to avoid the extremes of winter weather.

Neanderthals also displayed compassion and even spirituality. Archeologists have unearthed fossil remains of Neanderthals whose physical infirmities would have made them incapable of survival on their own—who would, in fact, have been a burden on their fellows. Earlier humans would surely have abandoned these aged or handicapped individuals, but the Neanderthals provided for them and found a place for them in their roving bands. The fossil record shows that when death came, Neanderthals laid the remains of their deceased tenderly to rest, larding the grave with foodstuffs and flint tools that presumably were provisions for an afterlife. In 1960, archeologists analyzed pollen traces collected from a Neanderthal burial site in Iraq and determined that the survivors had covered the grave with garlands of colorful flowers. There was a dark side to Neanderthal life; archeologists have found evidence that the Neanderthals sometimes indulged in cannibalism and infanticide. But they were, in any case, far more advanced than their ancestors.

For all this ingenuity and striving, however, the Neanderthals were destined only to endure, not to prevail. Some 40,000 years ago, a new breed of human emerged. Known as Cro-Magnons, after the site in France where their remains were first found, the newcomers were thoroughly modern in appearance—and evidently in intelligence as well: They were able to fashion a variety of sophisticated bone, flint and wooden tools greatly surpassing what the Neanderthals achieved. Where and how these advanced humans first arose is unknown. Fossil evidence from the Middle East and elsewhere suggests that Cro-Magnons may have evolved from Neanderthal populations. But this evolutionary process does not seem to have occurred in Western Europe; apparently, the Cro-Magnons migrated there. For the Neanderthals, the results of this influx were disastrous. Only about 5,000 years after Cro-Magnons arrived in Europe—a breathtakingly short interval, given the languid pace of evolutionary change up to that time—the Neanderthals had vanished.

The sudden disappearance of the Neanderthals of Western Europe presents one of the Ice Age's greatest puzzles. For a time, many scientists believed that the more advanced Cro-Magnons simply wiped out their hapless predecessors, but there is no fossil evidence of such wholesale slaughter. Other authorities maintain that the two populations mingled and interbred, or that the relatively backward Neanderthals fled in the face of the encroaching Cro-Magnons, finally to die out in the remote regions where they had sought refuge.

There can be no doubt about the power of Cro-Magnon intelligence. Living in larger and more organized groups than had earlier humans, Cro-Magnon peoples spread out until they populated most of the world. By 30,000 years ago they had reached Australia, having island-hopped by boat down the Malay Peninsula; by 25,000 years ago they had probably crossed the Bering land bridge and settled in Alaska, where they stayed until the melting of the Canadian ice sheets opened up a corridor through which they could continue southward.

The conquests of the Cro-Magnons coincided with the most bitter phase of the Ice Age. The ice sheets had begun a slow retreat about 60,000 years ago, although the climate remained severe. By 35,000 years ago the ice sheets were spreading once more from their frigid strongholds, creeping out like gargantuan amoebas and leaving lasting marks on the face of the earth.

Eighteen thousand years ago, when the sheets had reached their greatest advance, about one third of the earth's surface was buried under ice that averaged a mile in thickness. In the Northern Hemisphere, the largest sheet was centered over Hudson Bay. Called the Laurentide Ice Sheet, it coalesced with the smaller ice sheets of Ellesmere Island and Baffin Island to obliterate all of eastern Canada. Advancing southward at an average rate of between 200 and 400 feet a year, it bulldozed its way into New England, Illinois, Indiana and Ohio. Along its western margin, branches of the main flow merged with the Cordilleran Ice Sheet, which crept from the Canadian Rockies to smother parts of Alaska, much of western Canada, and parts of Washington, Idaho and Montana. The Greenland Ice Sheet grew by a third and linked up with the Ellesmere Ice Sheet; Iceland was, in fact, totally covered with ice.

Across the Atlantic, another ice sheet radiated out from the head of the Gulf of Bothnia in Scandinavia, reaching as far as the site of Moscow to the southeast, blanketing eastern Denmark and the northern sections of Germany and Poland, then merging in the North Sea with a smaller ice sheet that flowed from the Scottish highlands and the mountains of northern England, Wales and Ireland. Switzerland and neighboring parts of Austria, Germany, France and Italy were buried under a mass of Alpine ice that stretched from the Rhone above Lyon to Graz in Austria. The Pyrenees were an ice-armored barrier between France and Spain. Far to the north and east, another ice sheet developed on the northern Ural Mountains and the highlands of Novaya Zemlya. Small sheets formed in central and eastern Siberia and the Tien Shan, a range of mountains in central Asia, while the glaciers of the Himalayas extended far beyond their present boundaries.

In the Southern Hemisphere, the Antarctic ice sheet grew about 10 per cent larger than its present size, spreading seaward to the edge of the continental shelf. The floating ice shelves around its margins spawned icebergs prodigally, and the area of the ocean that froze over during the winter was greatly expanded. Worldwide, the spreading ice eventually covered half the surface area of the oceans. Mountain glaciers appeared in New Zealand, Tasmania and Australia. In South America, glaciers descended from the Andes and pushed far out onto the plains of Patagonia. Even the tropics felt the effects of the global chill; glaciers formed on Mauna Kea and Mauna Loa in Hawaii and Mount Elgon in Uganda—mountains that are today ice-free.

By the time the ice sheets had grown to their maximum volume—while waxing in some areas and waning in others—they had conquered about 17 million square miles of land. With more than 17 million cubic miles of water locked up as ice—almost three times the amount contained in today's ice sheets—sea levels fell dramatically. So much water had gone to nourish the ice sheets that, by the simplest calculation, the sea should have fallen by more than 500 feet.

In fact, however, the drop was offset by other forces set in motion by the

24

A New Look for Neanderthals

Neanderthals—the humans who occupied Europe between about 100,000 B.C. and 40,000 B.C.—have been victims of a profound misunderstanding. The appearance of the first Neanderthal fossils in the 19th Century coincided with public furor over Charles Darwin's theory of evolution. So distasteful was the notion that humans had primitive ancestors that the bones were dismissed as belonging to some sort of ape rather than to the genus Homo. Such ostracism received a ringing endorsement in 1913 when a prominent French scientist, Marcellin Boule, declared that Neanderthals were stoop-shouldered and bowlegged; he based these conclusions on a wildly faulty reconstruction of a single Neanderthal skeleton (one that—it was later pointed out—bears signs of severe arthritis).

As a result of new techniques for evaluating fossils, Neanderthals are now thought to have been an erect, agile breed whose intelligence was probably close to the modern level. This view is reflected in the paintings of Jay Matternes, an artist and naturalist who specializes in scientifically accurate renderings of primates and early humans. By means of a painstaking progression of sculptures and sketches (*right, and following pages*), Matternes has constructed layer after layer of muscle, fat and tissue to achieve a distillation of the best scientific thinking about a people who clearly figure in our ancestry.

A 19th Century drawing typifies the then-prevalent concept of a brutish Neanderthal. Scientific artist Jay Matternes, shown examining a skull cast at right, has developed a more accurate portrait (*pages 26-27*).

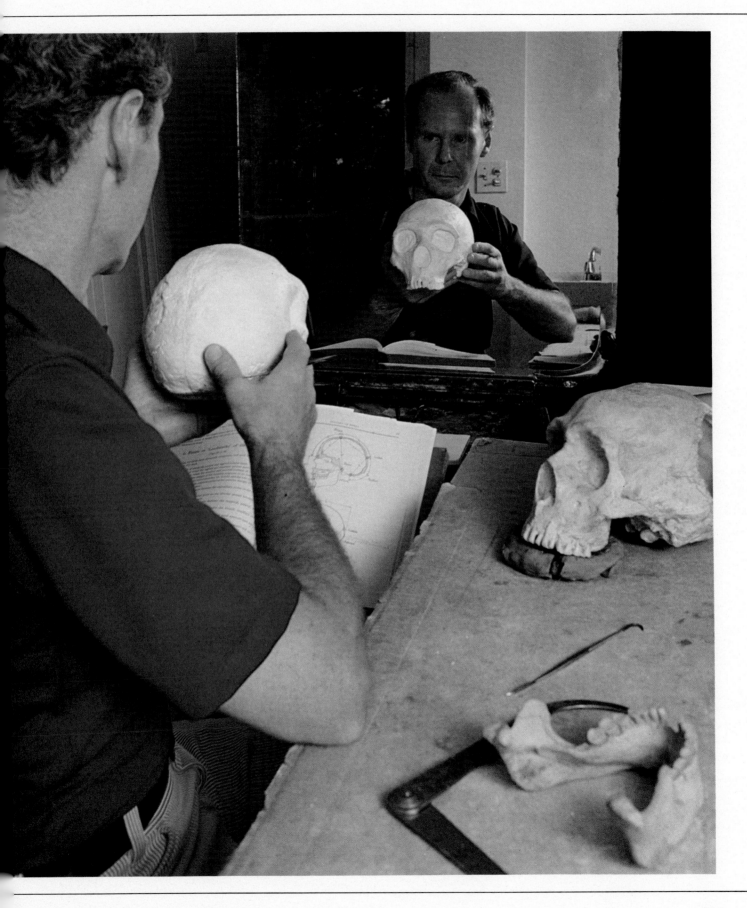

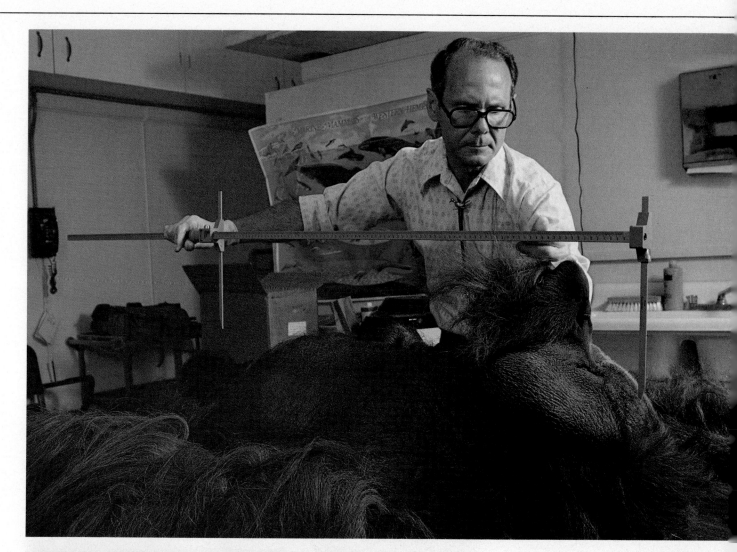

Careful study of an orangutan corpse has provided artist Jay Matternes with important anatomical information for his Neanderthal sketches. Surgically peeling back layers of skin, fat and muscle, Matternes measured and photographed the ape's tissue structure, which is similar to that of a human being.

Building carefully on his drawing of a Neanderthal skeleton, Matternes sketches muscle and fat. Striations on skeletons indicate areas where muscle was attached.

A Neanderthal male, shown without hair because the meticulous Matternes was unwilling to speculate where no evidence exists, emerges in distinctly human form.

ice. The sheer weight of the glacial masses deformed the surface of the earth, pushing the rigid crust into the elastic mantle by as much as a third of the thickness of the ice sheets. The sagging beneath the ice produced a corresponding upthrust on the ocean floors, which rose even farther as the weight of water on them was reduced. The ice sheets were so massive that they exerted a gravitational force on the seas, pulling the waters upward by perhaps as much as 100 feet near the margins of the ice.

When all these forces had balanced themselves out, sea level had fallen by about 400 feet, widening some coastal plains by as much as 250 miles and increasing the earth's land area by about 8 per cent. The Persian Gulf dried up; the Adriatic extended only about three quarters up the boot of Italy, putting the site of Venice 150 miles inland. The Ice Age created geographical unity where later there would be national divisions. Ireland was connected with Britain, which in turn was linked with France. North America and Eurasia were joined in the Bering Strait region, where the lowering of the sea created a plain that was more than 1,000 miles wide from north to south. The archipelago of islands that stretches southeast from the Malay Peninsula became an unbroken landform linking Sumatra, Java and the Philippines with the Asian mainland. Only a narrow channel separated this newly formed land mass from the Celebes; a second channel separated the Celebes from New Guinea, which became a northern extension of Australia.

The ice sheets in their passing left spectacular scars on the land that they violated. They scraped away soil, polishing and gouging the bedrock and pulverizing boulders to the consistency of flour. Where they encountered outcrops of rock, they flowed over or around the obstacles, plucking up boulders and carrying them far from their source. They scooped out deep basins and hollowed and deepened valleys into characteristic U-shaped profiles. South of Lake Ontario, the Laurentide Ice Sheet deepened a spray of 11 basins that are now known as the Finger Lakes; the largest, Cayuga Lake, is almost 40 miles long and more than 400 feet deep. In Norway, the ice sawed into the mountainous coastline and excavated far below sea level, leaving fjords as much as 4,000 feet deep.

Although they tended to follow river valleys—the line of least resistance—the advancing ice sheets occasionally broke through ridges and dammed rivers. The rivers would then find new outlets and proceed to cut new valleys, radically altering the drainage system. When ice sheets dammed the outlets of river valleys, the effect could be somewhat different. Near the border of Montana and Idaho, for example, an arm of the Cordilleran Ice Sheet blocked the Clark Fork River valley, holding back the water until it formed a lake with an area of more than 3,000 square miles. When the glacial dam eventually gave way, an estimated 300 cubic miles of water burst out—probably within a matter of days. The roaring torrent greatly deepened the river valley and carried boulders that measured more than 35 feet in diameter for hundreds of miles across the Columbia Plateau.

Where the ice sheets halted, they unceremoniously dumped their loads of unsorted debris, or drift, creating hills up to 150 feet in height. These are known as end moraines. In places where they overrode previously deposited drift, the glaciers shaped the debris into swarms of drumlins— clusters of egg-shaped hillocks lying parallel to the direction of ice flow. Other topographical features were formed by streams flowing within the ice and near the edge of the sheet. Rushing meltwaters cut tunnels in which they laid down their loads of sand and gravel; when the ice finally retreated, the deposits were left as long, sinuous ridges, called eskers, which can be found today winding across the countryside like eccentric railway embankments.

With the drop in sea level, rivers began to rush farther and more rapidly from their upland headwaters, deepening their valleys and leaving their former flood plains high and dry. But they were not able to handle the volume of debris carried by the streams of meltwater that were discharged from the margins of the ice sheets; the upper reaches of many major rivers became choked with coarse gravel and other glacial sediments, through which dozens of channels meandered, constantly changing course. Finer material—silt and clay—was picked up by the wind and transported in dense clouds far from the ice sheets. Where it settled, it eventually formed beds of fine-grained, homogeneous sediment up to 500 feet thick. In time, these wind-blown deposits, called loess, an old German word for "loose" or "light," covered more than one million square miles of North America, Europe and Asia.

The winds at the edges of the ice sheets were remarkably strong and frigid. Cold masses of air only 400 to 600 feet thick flowed down the sloping surface of the ice like sheets of water. Funneling into the valleys, these winds gained strength, on occasion reaching velocities in excess of 200 miles per hour as they burst out onto the plains.

At the bitterest stage of the Ice Age, 18,000 years ago, temperatures had dropped markedly in the zone that included most of Europe and the northern half of the United States. In Britain, average temperatures were some 12° F. lower than present-day averages, making it as cold as northern Alaska is at the present time; in the American Midwest, temperatures were as much as 18° F. lower. The subtropics and tropics were less severely affected; temperatures in the subtropics were about 4° F. below current averages, while the equatorial rain-forest belt was probably no cooler than it is today.

Temperatures on the surface of the ocean showed a similar pattern. The North Atlantic off Newfoundland was 26° F. colder in the summer than it is today; waters in the mid-Atlantic were 18° F. colder than now. At the Equator, however, water temperatures dropped much less—between 1° and 6° F. The cooling was more marked in the North Atlantic than in the North Pacific, partly because the closing of the Bering Strait blocked the escape of warm Pacific currents into the Arctic Ocean, and partly because the warmer Atlantic currents were driven southward. The Gulf Stream, instead of flowing northeast from Florida toward Scandinavia, drifted eastward toward Africa. The sea was then as cold off Spain as it is today around Greenland.

Since glaciers are able to grow only when more snow accumulates in winter than melts in summer, it might be assumed that precipitation increased during the Ice Age. In fact, the opposite happened in many areas: The atmospheric cooling reduced both the rate of evaporation from the sea (much of which was, in any case, covered with ice) and the moisture-carrying capacity of the air, resulting in an estimated 20 per cent decrease in precipitation worldwide. The decrease must have been even more pronounced over land, since the lowering of sea level and the consequent enlargement of the land masses increased the area of dry, continental climatic zones. Rainfall in Britain, for example, was between one third and one half of present-day averages, making the island as arid as Morocco is today.

In other regions, however, local conditions caused an increase in the rainfall and this, coupled with the reduced rate of evaporation, created lakes where now there are deserts. Death Valley in California held a lake more than 500 feet deep; Lake Bonneville covered much of present-day Utah and parts of Nevada and Idaho to depths of 1,000 feet and had a surface area nearly equal to that of modern Lake Michigan. In Africa, Lake Chad grew almost to the size of the Caspian Sea—which rose so high at the peak of the Ice Age that its waters mingled with those of the Aral Sea, 380 miles to the east.

Plant life was altered considerably by the combined assault of cold and increased aridity, and the earth's vegetation belts not only shifted toward the Equator, becoming compressed in the process, but also changed their composition. Around the margins of the ice sheets was a zone of treeless tundra clothed with arctic mosses, sedges, lichens and heathers. Beneath the thin skin of vegetation, which was watered by summer melting, the land—20 per cent of the earth's total—was frozen. In some places the freezing was permanent, and this permafrost penetrated to depths of as much as 1,000 feet.

Strange landforms appeared on the permanently frozen tundra. Water seeping into the ground and then freezing created lens-shaped masses of ice that pushed up the overlying surface into domes—called pingos—that were as much as 150 feet high and 1,800 feet across. The cold was so intense that the ground itself contracted and split open, forming 30-foot-deep rifts in which wedges of ice collected; when the frost cracks joined, they produced polygonal patterns, called patterned ground, that extended across hundreds of square miles.

At the height of the last ice age, the European tundra belt covered southern Britain, northern France, and Germany and Poland. To the east, it merged into patchy forest; to the south, it passed into a belt of cold, arid steppe that stretched from the Atlantic coast of Brittany to eastern Siberia. Small stands of birch, poplar and oak managed to survive in sheltered parts of south-central France, but true forests were pushed south of the Alps and Pyrenees. Even at these lower latitudes, evergreen conifers were the dominant trees; deciduous woodlands were found no farther north than the islands of the western Mediterranean, the southern part of Greece, and the shores of the Black Sea (then a freshwater lake) and the Caspian Sea. North Africa was probably no wetter than it is today, but in the cooler climate, extensive forests of drought-resistant conifers and Mediterranean shrubs grew in Morocco, Algeria and Tunisia. The Sahara was even then a desert, albeit a smaller and slightly cooler desert than it is today.

The North American tundra belt, which today is nearly 1,000 miles wide, was pushed south and squeezed into a narrow corridor, 30 to 150 miles wide, at roughly the latitude of New York City. On the Atlantic seaboard, the tundra gave way to boreal (northern) forests of spruce and pine that extended southward to what is now South Carolina, where temperate mixed forest began to appear. West of the Appalachians, the

The Tortured Faces of Frozen Ground

Ten thousand years after the retreat of the midlatitude ice sheets, the polar cold still extends its numbing influence far afield. About 20 per cent of the world's land area remains permanently frozen—in some cases to depths of almost a mile.

This periglacial environment, or permafrost, exists where the mean annual temperature of the soil remains below freezing. Although some melting usually occurs on the surface each summer, the underlying soil never thaws completely, and the depth of the permafrost increases slightly each winter until geothermal heat from below stops its descent. The building of a 100-foot-thick layer of permafrost takes thousands of years to accomplish.

The periglacial environment extends across about half of Canada and the Soviet Union, 85 per cent of Alaska, parts of Scandinavia and China, and, of course, almost all of the exposed land of Greenland. Human settlement of these areas is made extraordinarily difficult and expensive by the instability of the permafrost. In places, bridges, buildings and pipelines must be erected on specially designed pilings, which may have to be refrigerated to avoid softening the frozen ground.

The land itself almost seems alive. As seen here and on the following pages, even the slight melting and refreezing experienced from season to season in the Arctic and sub-Arctic environments over the centuries causes massive disruption of the soil and the emergence of a whole array of bizarre landforms.

An ice wedge pierces the permafrost of Garry Island, in Canada's Northwest Territories, to a depth of about 20 feet. Beginning as a trickle of meltwater freezing in a crack in the ground, ice wedges grow by small annual increments until they riddle large areas of the permafrost.

The repeated freezing and thawing of this permafrost area in the
Northwest Territories has lacerated its surface into polygons—some
of them measuring 100 feet across.

A shattered rock testifies to the long-term power of small quantities of ice. Melted by the heat of the sun on the rock, trickles of water found their way into tiny crevices in the boulder, froze and expanded with devastating effect.

Mystical "fairy rings" of stone found in Greenland are the work of ice, not sprites. The alternating contraction and expansion caused by thawing and freezing gradually forced coarser material to the surface, then moved it outward into the circular arrangements.

Conical hills called pingos are often pushed up in permafrost by ice accumulating beneath the surface. Sometimes 150 feet tall and 1,800 feet in diameter, they may collapse into volcano-like shapes such as this one on Canada's Tuktoyaktuk Peninsula—when the ice melts.

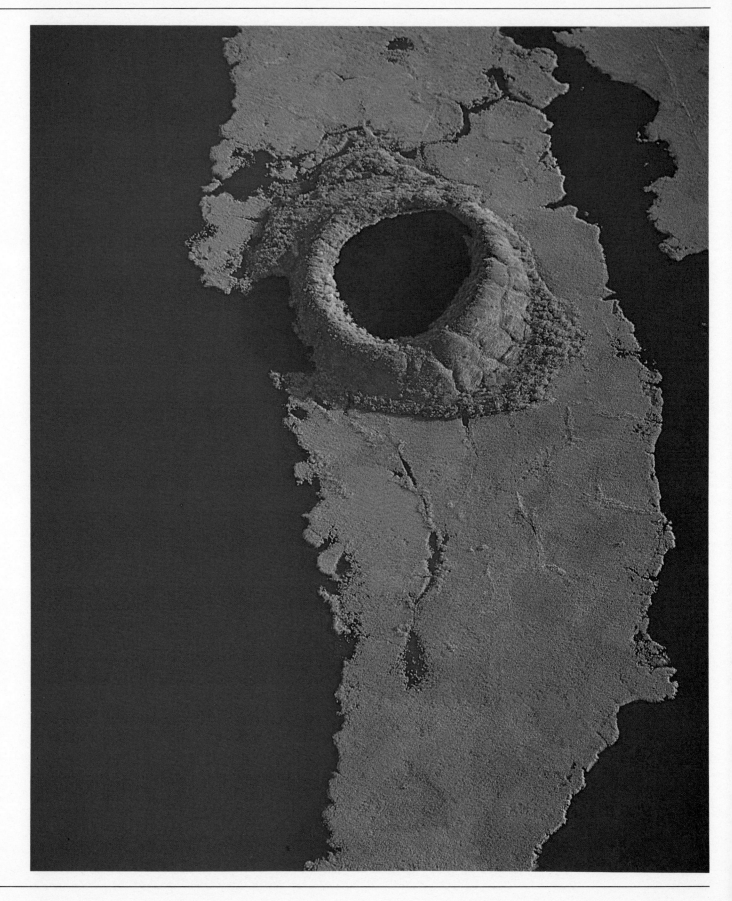

boreal forest, with birch, alder and tamarack mixed among the evergreen trees, spread to the Dakotas and invaded the prairies of the Midwest as far south as southern Iowa, northeastern Kansas and southwestern Missouri. The Great Plains were densely forested. In the southwest—one of the regions where rainfall increased during the Ice Age—the vegetation remained largely unchanged, although pine and spruce forests expanded at the expense of deciduous woodland and grassland in southwestern Nevada, and sagebrush and chaparral invaded large tracts of the Mojave and Sonoran deserts.

Soon after the ice sheets reached their maximum extent about 18,000 years ago they began to retreat. The waning was not uniform, and some advances continued to occur; but within about 5,000 years, the eastern margin of the Laurentide Ice Sheet had receded from the edge of the continental shelf to a position alongside the present coast of Maine. Farther west, the ice sheet still reached as far south as central Ohio, but just a thousand years later the United States was ice-free except for local icecaps in Maine and lobes in the Lake Superior and Lake Michigan basins. By 10,000 years ago, the ice sheets had withdrawn from the Great Lakes, the St. Lawrence estuary and parts of the Canadian Arctic islands, and had begun to retreat faster—in places at a rate of about 600 yards a year. By 7,000 years ago, the Laurentide Ice Sheet had split into two dying remnants, one on either side of Hudson Bay, and by 6,000 years ago, even these had disappeared. The Scandinavian Ice Sheet withdrew at about the same time as the North American ice sheets; the Danish archipelago and most of Britain were clear of ice by 14,000 years ago; and by 8,000 years ago, all that remained south of lat. 70° N. were a few disconnected icecaps in the Scandinavian mountains.

Through all of these titanic events, the Cro-Magnons showed astonishing resilience. Evidence gathered from archeological sites in Russia and France indicates some of their tactics for coping with the worst of the Ice Age. A group of Russian sites was discovered in the Kostienki-Bershevo area of the Don River valley, about 300 miles southeast of Moscow. Eighteen thousand years ago, the region was a cold, almost treeless steppe lacking caves or any other kind of natural sanctuary from the elements. The small bands of hunters who inhabited the area constructed their own shelters by digging pits a yard or so deep, ringing the excavations with the bones and tusks of mammoths (the main source of food), and then draping roofs of hide over the supports. To stave off the bitter cold, fires fed with mammoth bones were kept constantly burning.

By contrast, the Cro-Magnons of southwestern France lived for part of the year in the caves that honeycomb the limestone valleys of the Dordogne region. They, too, hunted mammoths, but their principal food was reindeer, which make up 99 per cent of the animal remains at many sites. Besides food, the reindeer supplied hides for clothing, antlers that were used as hammers to fashion flint tools and weapons, and bones that were carved into harpoons, awls, needles and fishhooks. Hunting required the ability to plan ahead, and the need to keep track of the seasonal movements of reindeer herds may have prompted the development of a means of keeping time: A reindeer antler discovered in France is engraved with a pattern that has been interpreted as indicating the phases of the moon.

Carefully crafted by Cro-Magnon artists some 15,000 years ago, a pair of two-foot-long clay bison rest against a limestone rock in a remote chamber of a

cave in Ariège, in southern France. The sculptured bison, which were discovered in 1912 by children, may have been used in ancient hunting rituals.

The paintings that adorn the walls of more than 200 caves in Europe are among the most impressive mementos of Ice Age people. Vibrant and realistic, most of the paintings depict prey animals—mammoths, horses, bison and deer—and predators such as lions and bears. Apart from their esthetic virtues, the paintings may have had a magical significance. Some of them are hidden away in deep recesses that could be reached only through tortuous tunnels. These images were obviously never intended to be admired as art. Perhaps the paintings were designed to somehow give the hunters power over their prey; if so, it is curious that portraits of the essential reindeer are far outnumbered by those of other animals.

There are few naturalistic representations of humans among either the cave paintings or other artifacts of the era. The people of the Ice Age may have believed, as do many tribes today, that a human portrait somehow captured the soul of its subject. Many of the human images are sculpted Venus figurines with exaggerated sexual organs; a number are enigmatic stick figures with animal features. There is another kind of painting, however, that strikes a very human chord. On the walls of some caves are outlines of hands made by pressing the hand to the wall and then spraying pigment over it, probably from the mouth or with a reed blowpipe. The handprints may have had ritual significance, for many of them show fingers that have been partially or completely amputated. But their immediate effect is to convey a sense of human individuality—as if the Ice Age artists were saying to future generations: "I exist."

The life of the Ice Age people was hard, but not brutally so. To be sure, it was a short life; infant mortality was high, and those men who survived until adulthood were lucky to reach the age of 40, while most women probably died in their 20s. But such things as cave paintings and sculpted

Ice Age hunters made cunning use of horseshoe-shaped prairie dunes like this one near Casper, Wyoming. They stampeded bison into the steep-walled sand formations, then slaughtered their trapped prey.

A bison skeleton emerges from its protective blanket of sand at a Casper, Wyoming, kill site after roughly 10,000 years. The scattered ribs in the foreground and a displaced foreleg to the right indicate that hunters had simply butchered the animal where it fell.

figurines—as well as flint tools that were finely crafted far beyond the point of mere utility—indicate that the Cro-Magnons had leisure time between the activities necessary for survival. Plant foods were plentiful in the warmer seasons, and there was a year-round bounty of game, both large and small. If the cave paintings do not evoke a golden age of human existence, they seem at least to speak of a world in which people lived in peace and harmony with nature.

But it was a short-lived world, and the climatic pendulum began to swing again from cold to the warmer conditions that would prevail through what we call modern times. Soon, the Cro-Magnons' way of life would change forever. Those who followed would learn to work with metals, to till the soil, to build cities and nations. And many of the great game animals that had sustained the Cro-Magnons and inspired their cave art would vanish from the earth. **Ω**

On desolate, sodden landscapes left by receding glaciers at Glacier Bay National Monument in Alaska, an age-old drama is being reenacted. The many stages of glacier retreat that coexist there show with remarkable clarity how vegetation reclaims barren land in a succession of plant communities, each dominated by species that ready the environment for their successors. Many scientists think the process resembles the slower reconquest of the land across North America and Europe after most of the continentwide glaciers of the Ice Age retreated 8,000 to 12,000 years ago.

Since about 1750, the glaciers that once blanketed the Glacier Bay region have been in retreat, exposing as much as 2,000 feet of terrain a year. For about a decade after the ice is gone, only mosses and a few stunted seedlings take root in the gravelly wastes. But after about 15 years, when rain and meltwater have leached alkaline minerals from the soil, the pace of revegetation quickens.

During the decades that follow, the soil nourishes a succession of shrubs, dense thickets, and sparse woodland. After 75 to 90 years, the soil is fertile enough to support forests, first spruce and then hemlock, which endure perhaps 1,000 years. In low-lying areas sphagnum moss may accumulate and choke the conifers, leaving a morass of bogs and dank ponds where, one day, a new forest may take root.

Glacier Bay

Gulf of Alaska

Rubble from a recently retreated glacier litters the landscape near the north end of Alaska's Glacier Bay. The area's glaciers, shown in white on the map above, have retreated as far as 62 miles from their maximum extent (blue tint).

ease with which their seeds and spores are carried by the wind, coupled with their natural hardiness, equips these plants for their role as pioneers.

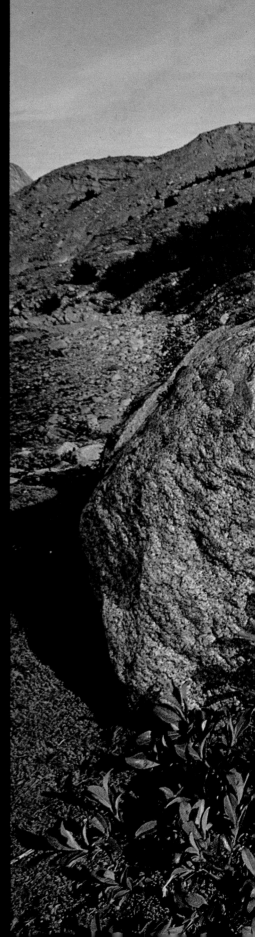

Alder seedlings sprout beside a boulder deposited by a retreating glacier about 25 years earlier. In another decade or so (provided that adequate soils are present), impassable alder thickets will dominate this landscape while the competing shrubs die out.

Disk-shaped Dryas mats dot a hillside that has been ice-free for perhaps 20 years. Bacteria living in the plant's roots enable it to change atmospheric nitrogen into soil nutrients that foster the growth of larger plants, such as the cottonwood saplings seen here.

Stands of cottonwood and young spruce flank a rubble-strewn landscape some 75 years after the ice's retreat. Unable to flourish until rain removed

alkaline deposits and earlier plant communities enriched the soil, the trees have now crowded out many of the shrubs and thickets that preceded them

Mature spruce forest cloaks the mountain
slopes near the southern end of Glacier
Bay, perhaps two centuries after the passing of
the ice. The success of the forest ensures its
eventual change; spruce seedlings cannot survive
in the shade of their parent trees.

The mossy remains of an alder thicket killed
by shade molder on the floor of a spruce forest.
Shade-tolerant hemlocks, growing in the rich
humus created by the litter of dead vegetation,
will replace the spruce within 600 years.

Decay overtakes the forest on the boggy fringes of a pond, where waterlogged moss has choked the roots of nearby trees. Such decay becomes widespread

1,000 to 1,500 years after postglacial vegetation first takes root. Yet the pond may fill in enough to support continued tree growth—a wetland forest

Chapter 2

THE MEGAFAUNA MYSTERY

The Siberian spring of 1846 arrived with freakish and devastating swiftness. In May of that year, according to a young man known to history only as Benkendorf, a Russian government survey team of which he was a member steamed from the East Siberian Sea into the mouth of the Indigirka River. The surveyors were bound for an inland rendezvous with a party of native guides, but they found the river in violent flood, swollen by melting snow and torrential rains, tearing away its banks and carrying great chunks of ice and frozen soil far out to sea. Abnormally high temperatures had sent meltwater cascading across the entire north coastal region and had also softened the surface of thousands of square miles of permafrost, turning the tundra into a treacherous morass and making an overland passage impossible. Clearly, survey work could not be carried out in these conditions, but the team nevertheless decided to venture upstream in their small steamboat. The only existing story of the journey, and the extraordinary discovery to which it led, was recorded by Benkendorf in a letter to a German friend.

"We steamed up the Indigirka on the first favorable day," wrote Benkendorf, "but we saw no signs of land. The landscape was flooded as far as the eye could see; we saw around us only a sea of dirty brown water, and knew that we were in the river only by the strength of the current. A lot of debris was coming downstream, uprooted trees, swamp litter and large masses of buoyant tufts of grass, so that navigation was not easy."

After toiling against the current for eight days, the surveyors reached the spot where they had arranged to meet the local guides. Not surprisingly, the guides failed to appear. "As we had been here in former years," Benkendorf wrote, "we knew the place. But how it had changed! The Indigirka, here about two miles wide, had torn up the land and made itself a fresh channel. When the floods subsided we saw, to our astonishment, that the old riverbed had become merely that of an insignificant stream. We went reconnoitering up the new stream, which had cut its way westward. Later, we landed on the new bank, and surveyed the undermining and destructive work of the wild waters that were carrying away, with extraordinary rapidity, masses of peat and loam.

"The stream was tearing away the soft sodden bank like chaff, so that it was dangerous to go near the brink. In a lull in conversation we heard, under our feet, a sudden gurgling and movement in the water under the bank. One of our men gave a shout, and pointed to a singular shapeless mass which was rising and falling in the swirling stream. I had noticed it, but had not paid it any attention, thinking it only driftwood. Now we all

Little changed from their ancestors of the Pleistocene epoch, two musk oxen take a characteristically defensive stance, effective against predators or the Arctic wind. The musk ox was one of relatively few large Ice Age mammals to survive the wholesale extinctions that occurred after the last glacial advance.

53

hastened to the bank. We had the boat brought up close, and waited until the mysterious thing should again show itself.

"Our patience was tried. At last, however, a huge black horrible mass bobbed up out of the water. We beheld a colossal elephant's head, armed with mighty tusks, its long trunk waving uncannily in the water, as though seeking something it had lost. Breathless with astonishment, I beheld the monster hardly 12 feet away, with the white of his half-open eyes showing. 'A mammoth! A mammoth!' someone shouted."

Fetching chains and ropes, the excited surveyors attempted to secure the carcass of the mammoth before the river carried it away, and after many tries they managed to get a line around its neck. Benkendorf then realized that the hindquarters of the beast were still embedded in the frozen river-bank, and he decided to let the river finish excavating the carcass before attempting to haul it onto dry land. A day passed before the beast was free of the ice, and during the wait the local guides arrived on horseback. With the aid of the horses and the extra men, the crew dragged the mammoth ashore and pulled it about 12 feet from the riverbank. For the first time, Benkendorf was able to get a good look at the animal.

"Picture to yourself an elephant with a body covered with thick fur," he wrote, "about 13 feet in height and 15 feet in length, with tusks eight feet long, thick and curving outwards at their ends. A stout trunk six feet long, colossal legs one and a half feet thick, and a tail bare up to the tip, which was covered with thick, tufty hair. The beast was fat and well-grown. Death had overwhelmed him in the fullness of his powers. His large, parchment-like, naked ears lay turned up over the head. About the shoulders and back he had stiff hair about a foot long, like a mane. The long outer hair was deep brown and coarsely rooted. The top of the head looked so wild and so steeped in mud that it resembled the ragged bark of an old oak. On the sides it was cleaner, and under the outer hair there appeared everywhere a wool, very soft, warm and thick, of a fallow brown tint. The giant was well protected against the cold."

Immediately after the mammoth came free from its frozen tomb, it had begun to decay. Benkendorf and the others tried to save as much of the carcass as they could. "First we hacked off the tusks and sent them aboard our boat," he wrote. "Then the natives tried to hew off the head, but this was slow work. As the belly of the brute was cut open, out rolled the intestines, and the stench was so dreadful that I could not avert my nausea and had to turn away. But I had the stomach cut out and dragged aside. It was well filled. The contents were instructive and well preserved. The chief contents were young shoots of fir and pine. A quantity of young fir cones, also in a chewed state, were mixed with the mass."

In their excitement, the men forgot the river and did not notice that the ground was slowly sinking beneath their feet. At length a cry of alarm from one of the Yakut guides distracted Benkendorf from the mammoth: "Startled, I sprang up and beheld how the undermined bank was caving in, to the imminent danger of our Yakuts and our laboriously rescued find. Fortunately our boat was close at hand, so that our natives were saved in the nick of time. But the carcass of the mammoth was swept away by the swift current and sank, never to appear to us again."

Benkendorf's dramatic find brought him face to face with one of the great mysteries of the Ice Age: What could have caused the sudden, worldwide

Kurtze
doch ausführliche
Beschreibung
Des
UNICORNU
FOSSILIS,
oder
gegrabenen
Einhorns/
Welches
in der Herrschafft Tonna
gefunden worden/
Verfertiget
von dem Collegio Medico
in Gotha/
den 14. Febr. 1696.
Daselbst gedruckt bey Christoph Reyhern/Fürstl.Sächß.
Hof-Buchdruckern.

extinction of the giant mammals, or megafauna, that inhabited the earth until only a few thousand years ago? After making a systematic review of past and present species, the eminent zoologist Alfred Russel Wallace would conclude several decades after Benkendorf's adventure: "We live in a zoologically impoverished world, from which all the hugest, and fiercest, and strangest forms have recently disappeared. It is surely a marvelous fact, and one that has hardly been sufficiently dwelt upon, this sudden dying out of so many large mammalia, not in one place only but over half the land surfaces of the globe."

The toll of death was especially apparent in Siberia, where, according to a 19th Century geologist, "the bones of elephants are said to be found occasionally crowded in heaps along the shores of the icy seas from Archangel to Behring's Straits, forming whole islands composed of bones and mud at the mouth of the Lena (a river west of the Indigirka), and encased in icebergs, from which they are melted out by the solar heat of their short summer in sufficient numbers to form an important item of commerce." More than 2,000 years ago, Chinese merchants were buying Siberian mammoth ivory for bowls, combs, knife handles and ornaments. From the descriptions of the natives, who had a superstitious dread of the great beasts that emerged each summer from their icy caskets, the Chinese gained the impression that the earth was the natural residence of mammoths. A book attributed to the 17th Century Chinese Emperor K'ang-hsi described the

In this 18th Century engraving, mammoth bones discovered in Germany are spectacularly misconstrued as a fossil of the legendary unicorn. Such attributions were not universally accepted. At the end of the previous century, claims of unicorn fossils had been debunked by a scholarly report whose title page is shown at top. Said the author: The alleged unicorn bones were merely mineral deposits that accidently resembled animal remains.

mammoth as "the underground rat of the north," which burrows through the ground with its huge teeth and "dies as soon as it comes into the air or is reached by sunlight."

In other areas of the world—where warmer temperatures precluded the preservation of body tissue—fossilized mammoth bones and teeth were found. For several centuries these enigmatic remains were seen as evidence of the existence of legendary giants. Around the year 1400, an English chronicler noted that there had been found on the Essex coast "two teeth of a certain giant of such a huge bigness that 200 such teeth as men have nowadays might be cut out of them." And in 1443 a "giant's thighbone" was unearthed in Vienna. Even after the discovery in the late 17th Century of two nearly complete mammoth skeletons in Belgium, the belief in prehistoric giants persisted, and it spread into the New World. When the fossilized tooth of a mastodon, a distant relative of the mammoth, was found in 1706 in a peat bog near Albany, New York, Governor Joseph Dudley of Massachusetts described the specimen in a letter to Cotton Mather, the Boston divine, pronouncing it the tooth of a giant killed in the Biblical Deluge. "Without doubt, he waded as long as he could keep his head above water, but must at length be confounded with all other creatures," Dudley explained to Mather, "and the new sediment after the flood gave him the depth we now find."

The myth of the antediluvian giants was not fully dispelled until 1806, when the woolly mammoth was scientifically classified by Johann Friedrich Blumenbach, a German expert on elephant fossils. After examining a number of mammoth bones found in Europe, Blumenbach concluded that they were an early form of elephant, which he named *Elephas primigenius*. When he later studied sketches of a well-preserved carcass that had been found in Siberia's Lena River, he connected the European fossils with the strange frozen creatures of the tundra.

One of the first people to grapple with some of the difficult questions surrounding the death of the gigantic Ice Age beasts was the Reverend William Buckland, professor of geology and mineralogy at Oxford University, who examined what he called a "charnel house" of prehistoric remains in a cave at Kirkdale, Yorkshire, in 1821. In addition to the fossils of animals native to cold and temperate climates, the cave contained the teeth and bones of warm-climate elephants, rhinoceroses and hippopotamuses. Scientists now know that these species inhabited northern Europe during the last interglacial, but Buckland was puzzled, he wrote in 1824, to find that a cave in northern England should be the last resting place of species that "at present exist only in tropical climates, and chiefly south of the equator." He was equally disturbed by the fact that the tropical species had apparently occupied the site only a short time after the cold-climate animals. However, he deliberately avoided drawing the conclusion that shifts in climatic conditions could have caused successive waves of extinction. After all, he explained lamely, "it is not essential to the point before me to find a solution."

In 1876 Alfred Wallace advanced one of the first and most obvious theories to explain the extinctions. Enough bones and fossils had been collected by then to permit him to catalogue and mourn the passing of a long list of vanished fauna: "In Europe, the great Irish elk, the *Machairodus* (saber-toothed cat) and cave lion, the rhinoceros, hippopotamus and elephant; in

North America, equally large felines, horses and tapirs larger than any now living, a llama as large as a camel, great mastodons and elephants, and an abundance of large megatheroids (ground sloths) of almost equal size." The elimination of so many species, Wallace wrote, must have been the result of some exceptional event that had occurred almost simultaneously in many parts of the world. Citing evidence that the northern portions of both Europe and North America had been covered with ice when these large animals were disappearing, Wallace maintained that the ice had probably "acted in various ways to have produced alterations of level of the ocean as well as vast local floods, which would have combined with the excessive cold to destroy animal life."

The mystery seemed to have been solved, although there were some scientists who questioned whether changes in climate could account for so many extinctions. And as the picture of the Ice Age grew more detailed during the next century, the climate hypothesis began to seem more and more inadequate. Although a sudden plunge into cold conditions certainly could have caused the deaths of the elephants and other tropical species that had been found by Buckland, it could not account for the extinction of animals that were well adapted to glacial conditions, such as the woolly mammoth and the woolly rhinoceros.

Some scientists suggested that there had been two waves of extinctions, with warm-climate species dying off as the glaciers advanced and cold-adapted species perishing when the ice sheets retreated. But if that were the case, others asked, how had the doomed animals survived through previous ice ages and interglacials? If the animals had been unable to cope with the pendulum swing of climate from warm to cold and back to warm, the fossil record should have shown waves of extinctions of approximately equal magnitude during each ice age. In fact, more species died out in Africa and North America during the last ice age than had vanished during all previous Pleistocene ice ages combined. And whereas the life-forms that had been lost during the earlier ice ages had been replaced by related species, the ecological niches that were emptied by the last wave of extinctions remained empty. In North America, for example, the mammoths, horses, camels, ground sloths, peccaries and giant beavers that once roamed the Great Plains disappeared virtually without replacement. (Horses did not reappear on the continent until they were brought to the New World by Spanish settlers in the 16th Century.)

Another element in the mystery was the curious selectivity of the extinctions. While a few species of small animals and birds became extinct, by far the greatest proportion of the species that disappeared were megafauna—those with an adult weight of more than 100 pounds. And the descendants of the large-animal species that did survive were smaller than their ancestors. The North American bison of today, for example, is the smallest member of a long race of bison; the brown bear of Eurasia is less than half the size of its Ice Age ancestors.

But it was the timing of the extinctions that cast the most serious doubts on Wallace's hypothesis. Even in his lifetime, scientists knew that not all of the extinctions had occurred during the coldest phase of the last ice age. Charles Darwin, on his famous expedition to South America aboard the *Beagle* beginning in 1831, had found fossils of many extinct species in sediments laid down after the last glaciation. Darwin's finds were significant;

Fossil Treasures of the Tundra

The tundra, a treeless region extending across the northern reaches of Europe, Asia and North America, was home to many of the giant mammals of the Ice Age—mammoths, bison, horses, musk oxen, camels, caribou and other plant eaters, along with the cats, wolves and short-faced bears that preyed on them.

As is still the case today, the tundra soil was frozen solid for much of the year, but the modest warmth of the Arctic summer thawed it to a depth ranging from a few inches to a few feet. At that point the spongy, saturated turf would erupt with grasses, lichens, flowers, mosses and shrubs. For the few warm months, grazers found rich pasturage; during the rest of the year, they subsisted on leftover leaves, twigs, shoots and lichens hidden beneath the snow.

Over the centuries, the animals that died on the tundra were preserved by the extreme cold. When the surface soil softened each summer, it lost its grip on the permanently frozen earth below and slid down even the most gradual slope. This annual process of thawing and sliding accumulated a wealth of megafauna and plant fossils in the Arctic's riverbeds and bottom lands.

The fossils were first reported in the 19th Century by ivory traders in Siberia and Alaska. When paleontologists investigated, they found the bones and flesh of a bestiary previously undreamed of. Hydraulic gold mining in Alaska later uncovered far more fossils than scientists could have found on their own; in 1938 alone, the American Museum of Natural History in New York received more than 8,000 specimens. The decline of gold mining reduced the rate of finds, but the increasing industrial development of the Far North may soon revive the once-bounteous flow of fossils from the great tundra depository.

Thawed by the brief summer of the Far North, the shallow, delicate soil of the Arctic tundra creeps down a slope. During the Ice Age, such movement often covered the bodies of fallen animals, preserving them for thousands of years.

A high-pressure jet of water cuts away the silt of the Alaskan tundra as miners search for gold. Often, they find ancient bones instead.

A paleontologist examines the preserved flesh of a bison that lived more than 30,000 years ago. Such specimens allow scientists to test their educated guesses—based on fossil bones alone—about the anatomy of extinct animals.

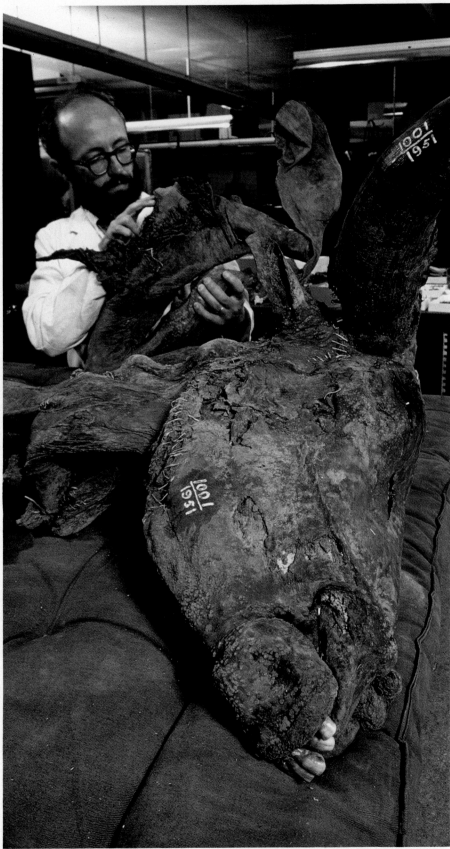

modern radiocarbon dating of the remains later confirmed that the main wave of extinctions occurred not when the ice sheets reached their greatest extent, but after they had begun to retreat. And so the mystery deepened.

Like a paleolithic detective story, the mystery of the Ice Age extinctions is strewn with bodies and contradictory clues. One way to identify the villain responsible for such an unprecedented toll of death is to hew to a venerable tradition of detective fiction and reconstruct the scene of the crime.

Scientists were aided in doing just that by an important fossil find made in 1901. That year, the prominent Los Angeles geologist W. W. Orcutt was asked to take a look at a strange collection of bones found by a well-drilling crew. The well site was near a natural asphalt deposit, where tar had been seeping up for centuries from deep, oil-bearing strata through beds of sand, silt and clay. Bones had risen to the surface occasionally over the years; they were assumed to be those of farm animals that had become trapped in the tar. But the well's owners were concerned about the number of bones that had spewed to the surface during the drilling. Orcutt examined what had been collected and quickly recognized the remains of a hyena-like dire wolf with a great head and stocky limbs, a lion-sized saber-toothed cat with heavy, tusklike teeth, and a giant ground sloth —all creatures that had inhabited the area more than 10,000 years before.

A tableau painted by Charles R. Knight shows why the tar pits at California's Rancho La Brea contain such a wealth of Ice Age fossils. Animals struggling to escape from the tar attracted predators, and these, in turn, were dragged down to their doom.

Since then, the tar pits of Rancho La Brea (later acquired by Los Angeles County and now surrounded by the city) have disgorged the bones of a remarkable compendium of Ice Age creatures—some 200 different species. Scientists have identified the bones of more than 1,000 saber-toothed cats and 1,600 dire wolves. Clearly, these animals had become mired in the tar, which probably was hidden beneath a pond. Old, sick or weakened predators, too slow to catch fleet-footed prey, probably waited near the pits to attack trapped animals, only to be sucked into the mire themselves. Other tar pits in Peru and the Soviet Union have turned up similar fossil bounty.

During the same year in which Orcutt made his remarkable discovery, a French anthropologist inched his way along the narrow entrance to a cave in southern France. At the end of the 200-yard passageway, Abbé Henri Breuil found a vaulted room etched with carvings of reindeer, lions, bears and—most significant—woolly mammoths. Until his discovery of Les Combarelles Cave, it had been assumed that the mammoth had disappeared from the earth before there were any human beings to observe the great beast, let alone make a record of its appearance. The few cave depictions of mammoths that had been found before Breuil's discovery had generally been regarded as fakes. But Breuil's stature as a scientist was such that, when he proclaimed that prehistoric human beings had indeed coexisted with the mammoth, he was able to convince the skeptics. Before long, cave

Paleontology students dip for fossils in the syrupy black mix of asphalt and mud that characterizes the Rancho La Brea site in downtown Los Angeles. Famous for spectacular animal remains, the La Brea pits also contain many plant fossils from the meadowland habitat of the animals that died there.

art of astonishing beauty and detail, presenting Ice Age beasts as eyes had seen them 20,000 years before, was being studied fervently in scores of European caves.

Many animals other than humans lived in caves and consumed their prey there, leaving behind extensive documentation of prehistoric dining habits. During the 19th Century, examination of a cave in South Devon unearthed fossilized bones of some 20,000 hyenas. The same cave was further excavated in the 1940s; this time, examination of cave-floor strata revealed that during a period of 200,000 years the cave had been occupied by successive generations of wolverines, bears, hippopotamuses, woolly rhinoceroses, elk and humans.

Occasionally, caves reveal curiously expressive traces of prehistoric creatures—the imprint of a jaguar paw on the wall of a cave in Tennessee where the large beast was apparently trapped (its bones were found nearby) or the deep scratches left high on the rock walls of a number of European caves by gigantic cave bears sharpening their powerful claws. Cave bears, in fact, account for the vast majority of fossils found in European caves. Though twice the size of a present-day brown bear and surely a fearsome sight, this gentle giant was essentially a herbivore. By thus piecing together clues from the fossil record, scientists have assembled a remarkably detailed picture of the great beasts of the Ice Age and the conditions in which they died.

The Pleistocene ice epoch had brought to a close the golden age of mammals, a period of cooling but relatively stable climate that had lasted more than 60 million years. When the warm-climate creatures of the temperate latitudes had felt the cold breath of the first glacial advance, they had begun to move south. In North America, where the mountain chains are aligned north to south, migrating animals had met few obstacles, and such mammals as camels and mastodons found suitable environs closer to the Equator. But in Europe, escape routes were blocked by the Alps and the Pyrenees. Trapped in an icy pincer, forced to compete for food in dwindling habitats, the European populations of warm-climate species such as the hippopotamus and lion had slowly perished.

Far from an extinction, this was a local and temporary tragedy. The glaciers retreated and the climate warmed periodically, and in a later interglacial about 130,000 years ago, hippos and rhinos were once again basking in the Thames, lions were ranging in the area of present-day Trafalgar Square and straight-tusked elephants were foraging in the woods. This repeated whipsawing of the climate between ice age and interglacial placed enormous evolutionary pressures on all living things. Because a large body retains heat well, the recurring cold of the Pleistocene encouraged the survival of the biggest. As a result, when the last glaciation began, the earth was host to a more varied and exotic bestiary than had ever existed before or would be seen again.

This lost world of giants included a North American beaver as large as a black bear, an armadillo from South America almost seven feet long and a 10-foot-tall Australian kangaroo. The cave lion, which ranged throughout Europe, was about 25 per cent larger than its modern relative. In North America, bison and horses shared the sweeping plains of the West with mammoths, lions, jaguars, tapirs, the fearsome plains cat and the fleet-footed short-faced bear. A living tank called the glyptodont hauled its

ILLUMINATOR OF A LOST WORLD

Until the end of the 19th Century, most museum fossil collections were kept in storage, accessible only to scholars. The few attempts at depicting extinct animals were a mix of shrewd guesses and flights of fancy.

Then a young New York artist with an avid interest in animals revolutionized the conception of the great creatures of prehistory. Charles R. Knight began his career in the 1890s by painting animals for stained-glass windows. Soon, however, a growing interest in portraying extinct animals prompted his collaboration with some of the leading American paleontologists of his day. Combining their scientific knowledge with his own observations of the habits and postures of living animals, Knight strove always to infuse his creations with life and movement.

By the 1900s, Knight's accomplishments were such that museums around the world were putting up exhibits of extinct animals, using his work as a guide to their reconstructions. Sculpting and painting for American museums during a career of almost five decades, Knight gave the world a vision of Ice Age animals unchallenged to this day.

A 1907 study by Charles R. Knight suggests how the undistinguished snout of Moeritherium *(upper right, above),* one of the earliest known ancestors of the modern elephant, might have evolved into the trunk and tusks of Elephas. Knight carefully charted the changes in the developing nose, lips and tusks, which are labeled on these three drawings.

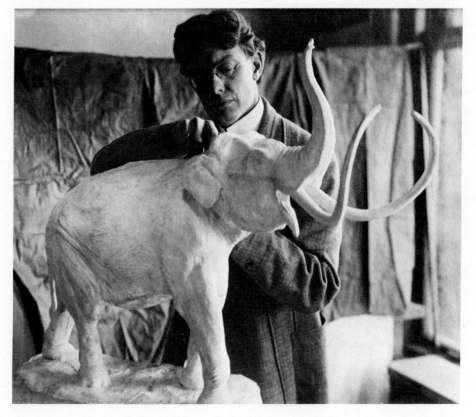

In this 1916 photograph, Knight works on a model of an imperial mammoth for the American Museum of Natural History in New York. He often sculpted the musculature of extinct animals before starting to paint them.

Woolly mammoths traverse a Pleistocene landscape in this Knight mural. The mammoth ranged the whole Northern Hemisphere, the woolly rhinoceros *(right)* only Eurasia.

From a high ledge, cave bears survey their territory. The huge but harmless bears (they were vegetarians) depended on caves for winter refuge and probably competed with humans for the natural shelters.

Knight's rendering of the Pleistocene Irish elk, Megaloceros, pictures a thickly muscled body—necessary to support its massive antlers, which spread nine feet or more.

Knight's 1903 depiction of the great Ice Age cat, Smilodon—muscles tensed and jaws agape—set new standards of realism and drama for reconstructions of extinct predators.

seven-foot-long armored bulk through the river valleys of Florida and Texas. Towering over them all, however, was the astonishing North American giant ground sloth, which weighed several tons and stood 20 feet tall as it balanced on its 36-inch-long feet and enormous tail to forage for leaves among the limbs of trees. Overhead soared enormous vultures with a wingspread of nearly 12 feet.

During the interglacial warming periods, the evolutionary trends were sometimes reversed, and increased body size became a liability. Usually this occurred where animals had colonized peninsulas during periods of maximum glaciation and lowered sea levels, only to be trapped on an island when melting glaciers flooded the seas again. The increased competition for available food supplies favored the survival of the occasional dwarfs produced by all species. Given the absence of large predators, and with enough time, entire races of dwarfs evolved; miniature elephants and hippopotamuses populated islands in the Mediterranean, tiny mammoths flourished on the Channel Islands off southern California, and even the huge ground sloths of tropical America evolved into dwarf forms on the Greater Antilles.

But more than any other species, the woolly mammoth has come to be regarded as the quintessential Ice Age beast. Its shaggy coat of black hair and a layer of blubber three inches thick provided insulation against −50° F. temperatures. The beast was further inured to frostbite by the fact that its sensitive ears and trunk were smaller than those of modern-day elephants. In summer, the mammoth fed on grass and other ground vegetation; in winter, it swept away snow with its enormous curved tusks to forage the grass underneath. The range of the woolly mammoth was extensive; herds migrated across the Bering land bridge some 600,000 years ago and joined other mammoth species that already populated most of North America. The largest, the imperial mammoth, which stood 14 feet at the shoulder and brandished tusks more than 12 feet long, roamed the southern Great Plains.

The distribution of fossilized bones reveals other extensive migrations of animals during the glaciation periods of the Pleistocene era. As if in partial compensation for the loss of habitats overrun by ice, the drop in sea level exposed both new areas for colonization and land bridges between continents. Lions and rhinos migrated from India and China into the chain of islands stretching southeast from mainland Asia. In the early Pleistocene, horses and camels migrated from North America across the Bering land bridge into Asia. During various glacial stages, mammoths, moose, musk oxen, bison, black bears and saber-toothed cats entered America from Asia. Migration probably stopped during interglacial stages—when the seas rose—then continued again as the ice sheets returned. Because the trip could be made only by species suited to the bitter cold of the Arctic route, hippopotamuses, for example, were denied the option. Other creatures, such as the woolly rhino, apparently never attempted the journey.

Migratory exchanges between North and South America enriched the populations of both. North American horses, cats, tapirs, llamas and peccaries descended upon South America in droves, in some cases thriving to the extent that they displaced native species. Among the emigrants from South America, porcupines and armadillos fared well in North America; but capybaras, glyptodonts and ground sloths were doomed by the harsher climate of the northern latitudes.

This, then, was the world in which the Ice Age megafauna died—a

world that had seen rapid and dramatic climatic change but that was, after the final glacial maximum, changing back to more or less the same conditions of climate and vegetation that had proved congenial to lions, sloths, camels and scores of other species during earlier interglacials. But something unprecedented was happening: The conclusion of the cold period was marked by a toll of death the likes of which had never before afflicted the creatures of the earth.

The first wave of extinctions hit Africa some 60,000 years ago, about the time when the last major glacial advance of the Pleistocene crested. During the next 20,000 years, 40 per cent of the large mammals on the continent disappeared, including giant baboons and pigs, antlered giraffes, long-horned buffalo, scimitar-toothed cats and three-toed horses. Eurasia was struck next. In Europe, about 50 per cent of the megafauna vanished, including the mammoth, the woolly rhino, the cave bear and the cave lion. The gradual elimination of these species also took about 20,000 years. In North America, however, the story was different. A full 70 per cent of the large animals—mastodons, mammoths, horses, camels, sloths, giant beavers, peccaries, dire wolves and countless others—died within the geological eyeblink of perhaps 1,000 years.

The evolution and ultimate extinction of species is a matter of course on earth. Variations of both flora and fauna tend to reach a maximum plateau of evolutionary success, then gradually decline and leave the stage to make room for new players. But the tempo of extinctions in North America was so accelerated—the decline of species that had occurred over some 2,000 generations in Africa and Europe was accomplished in North America in only 100 generations—that the life expectancy of many species was cut to an estimated 1/300 of their normal time on earth.

The first major scientific symposium on the mystery of the worldwide extinctions, sponsored by the National Academy of Sciences, took place in 1965 in Boulder, Colorado. Although some strikingly divergent opinions were aired, the participants tended to divide into three camps—one that attributed the deaths to climatic change, one that believed that prehistoric humans were responsible, and a third that laid the blame on a combination of these causes.

Among the advocates of the climate hypothesis was John Guilday, a paleontologist at the Carnegie Museum in Pittsburgh. According to Guilday, the major cause of the extinctions was the shrinking of the world's savanna and parkland zones, which were the natural habitats of the great herds of extinct herbivores—as they are today for the surviving game herds of Africa. Squeezed into a dwindling area of grassland by the encroaching forests and, in the American Southwest at least, by the growing deserts, the large herbivores such as camels, horses and sloths may have been forced into a competition so severe that few species survived. As a consequence, Guilday proposed, the large carnivores—lions, saber-toothed cats and short-faced bears—perished for lack of prey. The small mammals, however, were able to maintain viable populations, because they could subsist on less food and because they could find nourishment in nooks and crannies where the megafauna could not go. Among the large mammals that did survive the crisis, diminished ranges and territories encouraged dwarfing—the survival of smaller mutations.

A scuba-diving paleontologist inspects the leg bone of a mammoth on the bottom of the Aucilla River in northern Florida. Shallow during the animal's Ice Age lifetime, the river was deepened by a rise in sea level as the ice melted, and the water preserved the bones.

The crisis did seem to affect the grassland fauna more than the forest animals; in Europe, for example, all the herbivores that perished were denizens of the steppe and tundra, including species such as the giant elk, which could never have maneuvered its huge spread of antlers in dense forest. But other evidence—apart from the fact that the doomed herbivores and their predators had survived previous interglacials—indicated a basic flaw in the climate hypothesis. For one thing, the habitats of many of the extinct species actually increased in size. The point had been made in 1946 by F. C. Hibben, an anthropologist at the University of New Mexico: "Horses, camels, sloths and antelopes all found slim pickings in their former habitats. But what was to prevent these animals from simply following the retreating ice to find just the type of vegetation and just the climate they desired? If Newport is cold in the winter, go to Florida. If Washington becomes too hot in the summer, go to Maine." The woolly mammoth was one species that extended its range as the ice sheets retreated. In North America, the tundra expanded from a belt that was little more than 100 miles wide to a vast wilderness extending across much of Canada and Alaska.

However, according to some scientists, the habitat in which the mammoth met its end differed in another important respect from the Ice Age tundra on which it had thrived. As the glaciers retreated, the tundra zone moved much farther north, where the growing season is short, and precipitation became far more frequent than it had been on the dry, steppelike tundra of the Ice Age, when snowfalls must have been sufficiently light to enable the woolly mammoths to get at the underlying vegetation.

Curiously, the conditions in which Benkendorf found the mammoth in the Indigirka River may have been similar to those in which it died. The surveyor was surprised to find that the carcass was in an upright position, as if the ground on which the animal had stepped, "thousands of years ago, gave way under the weight of the giant, and he sank as he stood, on all four feet." Since that carcass was lost, the time of its death can only be guessed, but hundreds of other Siberian mammoths have been found in identical positions, suggesting that they perished when a rapid thaw melted the permafrost and turned the tundra into a huge bog. A succession of warm, wet springs could conceivably have wiped out most of the Siberian mammoths as they moved northward on their regular springtime migrations.

However, local fluctuations of climate do not account satisfactorily for the other extinctions. While trying to explain the global sweep of death, participants at the Boulder symposium offered some novel suggestions. Anthropologist William Ellis Edwards raised the possibility that while the Eurasian and American megafauna had been cut off from each other by the American ice sheets, they had developed and become tolerant to differing parasitic diseases; then, when deglaciation brought them into contact again, the animals on each continent had succumbed to the diseases of the other. Edwards then rejected the hypothesis, however, on the grounds that infectious disease declines with population, until scattered but highly resistant individuals are left.

Edwards also dismissed the notion that the megafauna may have fallen victim to racial senility—a rather abstract concept based on the idea that species living for a long time under optimal conditions, where the laws of natural selection are relaxed, produce inferior strains that would make them susceptible to sudden changes in their environment. The idea had been born

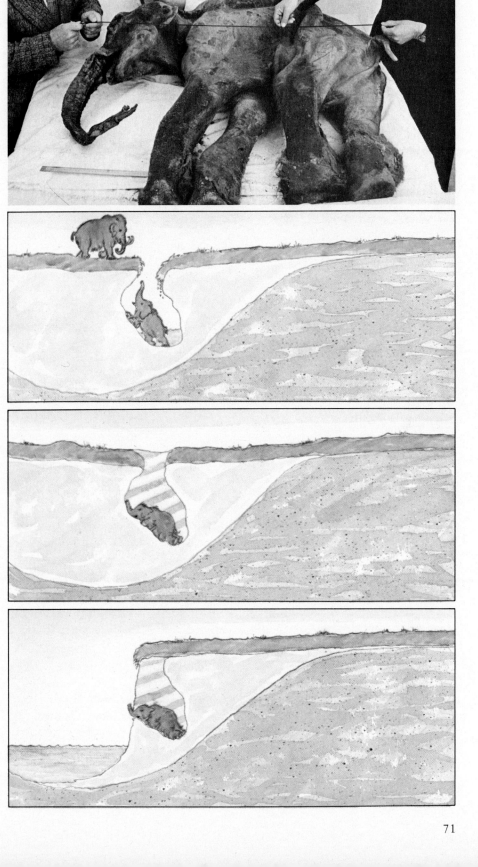

Soviet scientists examine the 40,000-year-old carcass of a woolly mammoth calf, found in 1977 in an eastern Siberia creek bed. The drawings below suggest the probable chain of events leading to the discovery.

Following its mother across the tundra, the calf falls through a thin layer of frozen turf into a channel cut by melting water in the underlying permafrost. The animal starved, then froze, and was soon entombed as deposits of ice and sediment filled the hole.

For thousands of years the mammoth lay hidden in the frozen soil, preserved not only by the extreme cold but also by a high level of tannic acid—a product of decayed vegetation—in the marshy deposits of its grave.

Eventually, the shifting channel of a river approaches the mammoth's tomb, cutting away the soil and exposing the carcass. Animals uncovered this way are often located by the foul odor that rises from their rapidly decomposing flesh as it thaws.

Fragile Survivors of Glacial Adversity

As the great ice sheets of the Pleistocene epoch spread across the Far North, they extinguished all life on the land that they covered. But here and there, improbable bits of vitality and color—tiny, frail butterflies and moths—flickered over the frozen wasteland. Flying from small, ice-free refuges on mountain peaks that crested the glaciers, or from indentations in sheltered cliff faces, these delicate creatures rode out the last glaciation, their larvae subsisting on the scant vegetation that shared their havens.

Today, such insects and plants are known as relicts—relatively small communities of species that differ markedly from the more recently established life-forms around them. Frequently, the relict colony is a remote outpost of a larger population whose ancestors retreated before the ice sheets and took up residence in more amenable locales farther to the south.

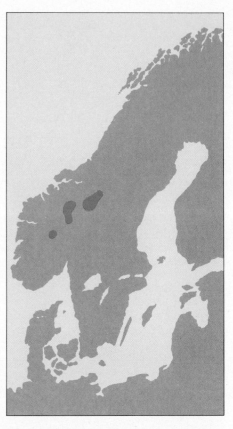

While the well-insulated, fur-covered animals fled the Ice Age or perished, a delicate butterfly found haven in the Far North. The butterfly is still found in the areas of Scandinavia shown in red on the map, near the Arctic Circle.

in Europe, where paleontologists were impressed by the fact that most of the cave-bear remains they found were of cubs and old individuals suffering from rickets, bone inflammations and gout so severe that some specimens had several bones growing together in one unwieldy structure. There was also evidence that many of these defects had been transmitted by inbreeding. The cave bear appeared to have been a species on its last evolutionary legs, but the paleontologist Björn Kurtén explained the disproportionate number of diseased individuals by pointing out that cave remains would inevitably include a majority of cubs and old, diseased specimens, since healthy, mature individuals would have died in the open of natural causes. As for the evidence of inbreeding, Kurtén interpreted it as the result, not the cause, of a decline in population.

After examining all the alternatives, many participants at the Boulder conference, including Edwards, accepted the idea that the real agent of the extinctions, the only factor unique to the last ice age and its unprecedented aftermath of death, was humankind. It was a view that Alfred Wallace, in a dramatic and influential change of mind, had come to espouse before his death in 1913. Still, the notion had remained distasteful to many, partly because of the lack of conclusive evidence and partly because of the persistent and romantic image of the Paleolithic hunter as a noble savage. The famous vertebrate paleontologist Wolfgang Soergel contributed to that image when he wrote in 1912, "In all the hunting that ends with the extermination of a species, the motivation is never hunger. Money, and the greed for it, have been the incentive. The savage does not know these; he hunts to eat and so is unable to decimate the big game to any important extent."

Paul Martin, a professor of geosciences at the University of Arizona and an organizer of the Boulder conference, was similarly reluctant to place the blame on humans. "The thought that prehistoric hunters 10,000 to 15,000 years ago (and in Africa over 40,000 years ago) exterminated far more large animals than has modern man with modern weapons and advanced technology," he wrote, "is certainly provocative and perhaps even deeply disturbing." Yet Martin, an expert in geochronology, nevertheless contended that the circumstantial evidence connecting the extinctions with the arrival of human beings was strong. Even in historical times, the appearance of primitive people has been accompanied by the disappearance of some animal species. In New Zealand, for example, 27 species of flightless bird—including a giant moa more than nine feet in height—died out shortly after the arrival of humans in the Ninth Century A.D.

Despite the primitive armaments of early hunters—stone knives, axes and spears—and the enormous populations of large, powerful animals that were obliterated, the evidence that tends to incriminate the noble savage is more than circumstantial. By the end of the Ice Age, humans definitely had developed techniques of wholesale slaughter that went far beyond satisfying immediate food requirements. At several locations around the world, paleontologists have reconstructed the circumstances in which large animals were stampeded into arroyos or stream channels, where they were trapped and killed; the uppermost animals were butchered, the rest were left untouched. Remains found at a Russian site indicate that during a single hunt nearly 1,000 bison were killed with at least 270 spears tipped with flint and 35 tipped with bone. Another mass hunting technique may have been the fire drive, in which deliberately set fires were used to stampede prey toward waiting hunters.

All animals—fierce carnivores as well as harmless grazers—are today so fearful of people that it is difficult to imagine a time in the fairly recent past when animals had no instinctive terror of human hunters. Yet if North America remained unpopulated until about 20,000 years ago, its fauna would have been particularly vulnerable to the newly arrived predator. Charles Darwin found a similar situation in the 1830s, when he arrived on the Galápagos Islands to find the animals so tame that foxes could be coaxed into the range of a knife, and birds could be killed with a hat or a walking stick. "We may infer," he wrote, "what havoc the introduction of a new beast of prey must cause in a country before the instincts of the indigenous inhabitants have become adapted to the stranger's craft or power."

Dwarfing can also be an indirect result of human predation. In shooting reserves and game parks, hunting for trophies or for meat invariably results in a reduction in the average size of prey species. When unusually small variants, or dwarfs, are born, they are more likely to survive because the pressures on the herd select for adults that mature and breed quickly. For an antelope confronted by a leopard, size and strength may make the difference between escape and capture; however, when the hunter is armed with projectiles, size may be more a handicap than a useful defense.

Ironically, Stone Age hunters may have inadvertently caused the extinction of some species by actively trying to conserve their main prey. According to Grover Krantz, an American anthropologist at Washington State University, early bison hunters may have avoided killing females and young and thus contributed to an increase in the total bison population and the elimination of horse, antelope, camel and other species that competed with the bison for the same food. This hypothesis also explains why many hunting peoples in historical times, such as the bison-hunting Indians of North America, did not exterminate their prey. In Europe the reindeer hunters may have tampered with the ecology in the same way, thus ensuring the reindeer's survival while contributing to the extinction of the mammoth, woolly rhinoceros and steppe bison that shared its habitat. From selective killing of reindeer, it must have been but a short step to active herd man-

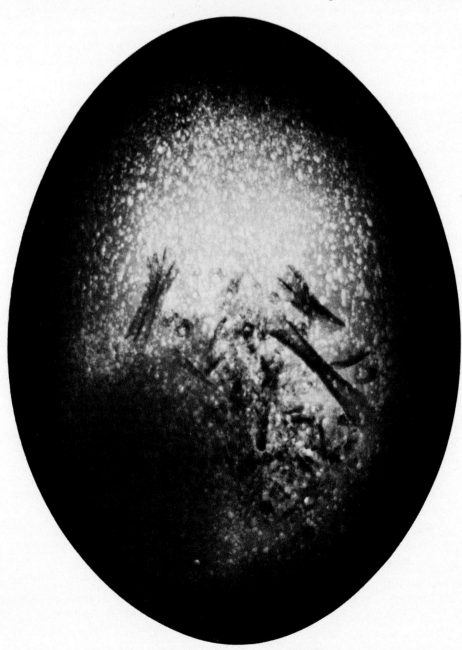

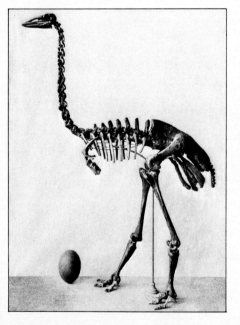

Fossils such as this watermelon-sized egg (X-rayed in order to show its embryo) and the skeleton above are all that remain of the 10-foot-tall Aepyornis of Madagascar. An island resident for millions of years, the flightless bird became extinct not long after the arrival of humans less than 1,000 years ago.

agement and the form of semidomestication practiced today by the Lapps of Scandinavia—a form of husbandry that is part hunting, because it involves pursuit, and part herding, because it involves guarding the animals as a private resource.

If Paleolithic hunters wittingly or unwittingly caused the extermination of the Ice Age megafauna, then most of the environmental niches of wilderness North America are artificially empty; Arizona, for example, should have camels, elephants and tapirs. According to Paul Martin, the loss of the indigenous species may have upset the ecosystem, making the North American wilderness susceptible to fire, insect invasion, brush expansion and erosion; he has suggested that one way to restore the ecological balance would be to introduce exotic animals of African or Asian origin as replacements for the original fauna.

In Scots law there is a verdict "not proven," which can be handed down in cases where insufficient evidence exists to warrant a "guilty" or "not guilty" verdict. The case against humans as the destroyer of the Ice Age megafauna has never been proved. In fact, during the years after the Boulder conference the pendulum of scientific consensus appeared to swing back toward the climatic theory of the extinctions—that environmental changes during the postglacial warming period altered the interrelationships among animals in a variety of ways, whose cumulative effect was the great dying out of the megafauna.

At a 1982 meeting of the Society of American Archeology held at the University of Minnesota, the causes of the extinctions were debated again by Paul Martin and vertebrate paleontologist Russell Graham. Arguing from a sizable body of recent research, Graham held that climatic changes destabilized the evolutionary equilibrium among species. "It's like a deck of cards," he contended; "after they are shuffled you wind up with a whole new arrangement," with animals thrown into competitive situations that did not previously exist. Humans may have caused the obliteration of a few species, Graham said, but could not have killed such enormous numbers of animals. Before the meeting was over, the several hundred archeologists and paleontologists present were asked by Martin who among them subscribed to the thesis of full human responsibility for the extinction of the megafauna. Only one person raised a hand. The overwhelming majority of those present subscribed to the belief that the extinctions were the result of a multiplicity of causes.

After all the decades of research, discovery and debate dedicated to finding out how and why the world had become, in Alfred Russel Wallace's words, "zoologically impoverished," it appears that the eminent British geologist Sir Charles Lyell was as close to the truth as anyone when he wrote in 1863: "It is probable that causes more general and powerful than the agency of Man, alterations in climate, variations in the range of many species of animals, vertebrate and invertebrate, and of plants, geographical changes in the height, depth, and extent of land and sea, or all of these combined, have given rise in a vast series of years to the annihilation of many large mammalia." Ω

SKELETON KEYS TO AN AGE OF GIANTS

The scale of life in the Pleistocene epoch is nowhere more dramatically conveyed than in the fossil remains of the Ice Age megafauna. Such skeletons as those on view at the Smithsonian Institution in Washington, D.C. *(right, and following pages)*, offer tangible proof of a world filled with massive creatures having no counterparts in present times.

The implicit power of these animals only deepens the mystery of their precipitous disappearance at the end of the last glaciation, about 10,000 years ago. Evidence suggests, however, that great size hindered the ability of some species to adapt to a climatic change.

The Irish elk *(pages 78-79)*, for example, is believed to have fallen victim to improving weather conditions. As the ice sheets retreated and temperatures rose, the grasslands that formed the Irish elk's habitat were overgrown by dense forests—an inhospitable environment for an animal with such a great spread of antlers.

Woolly mammoths, although apparently free to follow the northward retreat of their favored tundra grazing lands, also died out. The presence of flint spearheads among mammoth bones suggests that a dangerous new predator may have tipped the balance against them.

Scientists may never be able to put together a satisfactory explanation for the demise of some of these creatures. But the overall toll is unmistakably clear: More than half of all Pleistocene mammal species vanished from the earth.

The spectacular fangs of a saber-toothed cat, or Smilodon, attest to its killing power. Yet the animal's bones reveal that it was a heavy, slow hunter that would have preyed on such large and ungainly animals as ground sloths and tapirs.

The Irish elk, which ranged across northern Europe and Asia, grew 100-pound antlers—spanning nine feet in this specimen—every summer and shed them in winter. Although the creature was nearly extinct by the close of the last ice age, isolated colonies survived in Central Europe until about 500 B.C.

The jaws and teeth of these dire wolves
testify to the aptness of their name. Like modern
wolves, they hunted in packs; but they relied
more on a plodding, relentless pursuit than on
swiftness of either brain or foot.

The grasping claws and elephantine size of
the giant ground sloth give its skeleton
a ferocious appearance; in fact, the creature—
which roamed both North and South
America—was an inoffensive plant eater.

Despite the impression of great size conveyed by its enormous curved tusks, the woolly mammoth stood no taller than a large Asian elephant and had a shorter body.

Chapter 3

SOLVING THE "COSMIC PROBLEM"

According to Norse mythology, the world will end not by fire but by ice, formed during endless winters that will freeze the very seas. Other Scandinavian myths have it that the earth and the heavens were made from the body of a great frost giant who materialized out of mists rising from melting ice—ice that had once imprisoned the entire universe. Perhaps such tales represent ancient memories of the Ice Age; if so, they are among the few folkloric allusions to the glacial past. Although fully modern humans witnessed the waning of the most recent ice age, even the oldest legends contain no images of retreating glaciers or warming climates. People of many cultures tell each other of cataclysmic floods that once covered the world with water, but never speak of a time when towering ice sheets blanketed much of the earth.

Indeed, it was not until the 19th Century that scientists began to decipher the geological clues indicating that a great ice age had preceded the rise of human civilization, or to confront the daunting mystery of what could have caused such widespread glaciation. And in 1837, when the Swiss scientist Louis Agassiz first laid out a scenario of far-reaching ice-age conditions, his colleagues were, to say the least, highly skeptical.

On July 24 of that year, as Agassiz stepped forward to address the Swiss Society of Natural Sciences, of which he was president, he was greeted by enthusiastic applause. Although only 30 years old, Agassiz was already recognized as a leading authority on fossil fish, and this meeting—held in the Swiss town of Neuchâtel—was expected to confirm his reputation as one of the brightest luminaries of the dawning scientific age. Agassiz had recently examined a large number of fossil-fish specimens gathered by an expedition to Brazil, and the society members were eagerly looking forward to hearing about his findings; instead, he addressed an issue that lay much closer to home and that had already caused considerable controversy when it was raised earlier by two other members of the society, Ignatz Venetz and Jean de Charpentier. "I have in mind," said the young naturalist, "glaciers, moraines and erratic boulders."

As the assembled scientists stirred uneasily, Agassiz pointed out that Switzerland's Jura Mountains are littered with granite boulders completely unlike the limestone on which they rest. He argued that these boulders, called erratics, must have been transported to their locations by glaciers, and he backed up this claim by drawing attention to the scratches and striations that scarred much of the exposed bedrock in the Jura—evidence, he said, that the glaciers still dotting the region had once extended far

Stranded high on the side of a hill by the withdrawal of a glacier some 13,000 years ago, a huge boulder dominates an ice-carved landscape in Maine's Acadia National Park.

Swiss-American naturalist Louis Agassiz *(left)* listens as his colleague at Harvard University, mathematician Benjamin Peirce, makes a geographical point in a photograph taken around 1871. Agassiz was the most influential of the early scientists who proposed that much of the topography of northern Europe was shaped by widespread glaciation—the Ice Age.

beyond their present limits. Warming to his theme, Agassiz presented the startled audience with a vision of a distant age when ice sheets stretched from the North Pole to the shores of the Mediterranean and Caspian Seas, and he continued by saying that such a cataclysmic event would have annihilated many of the earth's creatures. The next day, Agassiz appropriated a term coined a year before by a botanist friend, Karl Schimper, and called this global trauma the *Eiszeit*—the Ice Age.

Eventually, Agassiz's address would be widely known as the "Discourse of Neuchâtel" and would be recognized as a landmark in the annals of science. At the time, however, his remarks were greeted by a stunned, uncomprehending silence. Then furor broke out among the scientists. Few of them quarreled with Agassiz's assertion that the planet had undergone catastrophes in the past; in the early 19th Century, the phenomena of massively folded strata, uplifted marine sediments and fossils of extinct creatures were interpreted as evidence of geological forces powerful enough to warp the surface of the planet and to annihilate life. But few in the audience could imagine such an apocalyptic role for ice.

According to the conventional wisdom of the time, the widespread evidence of geological trauma simply authenticated the Biblical narrative of the Great Flood of Noah; glacial deposits of clay and gravel, for example, were classified as diluvial, or floodborne, and were referred to as drift. Admittedly, it was difficult to explain how the huge erratic boulders mentioned by Agassiz could have been transported many miles by water. The Scottish geologist Charles Lyell had made a commendable attempt in 1830 by proposing that the boulders had been frozen into the base of icebergs drifting on the waters of the Flood. This notion seemed more than a little strained—but it was less blasphemous than what Agassiz was proposing.

Agassiz was hardly the first person to challenge the doctrine that geological devastation could be explained in terms of the Flood. The idea that the granite erratics of the Jura had been carried by glaciers had been advanced as early as 1787 by a Swiss lawyer named Bernard Friedrich Kuhn. And in 1795, the same conclusion had been reached by James Hutton, the Scot regarded as "the father of geology." Hutton's law of uniformitarianism—that all the physical features of the earth can be explained in terms of natural processes working at a more or less uniform rate—was both the cornerstone of geological science and a refutation of the theory of catastrophism. In his own time, however, Hutton's work was largely ignored, partly because his law of uniformitarianism implied that the earth was much older than his contemporaries were prepared to believe. For Hutton's generation, little had changed since 1654, when John Lightfoot, the Vice-chancellor of Cambridge University, had assigned a date and time to Archbishop James Ussher's earlier calculation of the earth's year of origin. Declared Lightfoot: "Heaven and Earth and clouds full of water and Man were created by the Trinity on 26th October 4004 B.C. at nine o'clock in the morning."

Ironically, the suggestions about the geological impact of glaciers that the learned scientists of the world found to be unacceptably radical had for years been self-evident to the natives of the Swiss Alps. From their own observations, they were aware that glaciers not only moved but also shaped the land in their passing. A guide and chamois hunter from the southern Swiss Alps, Jean Pierre Perraudin, had written in 1818: "Having long ago observed marks or scars occurring on hard rocks which do not weather, these marks being always in the direction of the valleys, I finally decided, after going near the glaciers, that they had been made by the pressure or weight of these masses, of which I find traces at least as far as Champsec. This makes me think that glaciers filled in the past the entire Val de Bagnes, and I am ready to demonstrate this fact to incredulous people by the obvious proof of comparing these marks with those uncovered by glaciers at present."

Ignatz Venetz, a Swiss civil engineer, heard of Perraudin's ideas, and was so impressed that he devoted a substantial amount of time over a number of years to investigating glaciers in various parts of the Swiss Alps. By 1829, he had amassed enough information to present his conclusions to the annual meeting of the Swiss Society of Natural Sciences, and a skeptical audience heard Venetz maintain that Alpine glaciers had once extended not only over the Jura but also northward into the European plain. His only sympathetic listener was Jean de Charpentier, an eminent scientist who was director of the salt mines in the Swiss town of Bex. Earlier, de Charpentier had dismissed Perraudin's original claims as extravagant, but now he enthusiastically supported the scholarly Venetz.

Three years later, a German scientist named Reinhard Bernhardi published a paper in which he made the startling claim that polar ice had once reached as far as southern Germany, littering it with erratic boulders. "This ice," wrote Bernhardi, "in the course of thousands of years, shrank to its present proportions, and the deposits of erratics must be identified with the walls or mounds of rock fragments which are deposited by glaciers large and small, or in other words are nothing less than the moraines which this vast sea of ice deposited in its shrinkage and retreat." This was the most succinct statement about a former ice age yet to appear in print, but it attracted little interest, and its author—now regarded by many as one of the true originators of the ice age idea—was relegated to the sidelines while others took up the cause.

In 1834, Jean de Charpentier outlined for the Swiss Society of Natural Sciences some of the geological evidence supporting Ignatz Venetz' claims. Louis Agassiz was among those who heard de Charpentier's presentation, but the young naturalist was singularly unimpressed. Two years later, though, he spent the summer in de Charpentier's hometown—the resort of Bex—and when he saw the evidence with his own eyes, he embraced the glacial theory with the fervor of a religious convert. Agassiz went further— extending his colleagues' research and, with visionary zeal, using it as the basis of his proposal that an ancient ice age had affected the whole world.

When he gave his famous discourse in 1837, Agassiz felt he had marshaled irrefutable evidence, but his soaring imagination was brought down to earth by a barrage of criticism. Even de Charpentier was shocked to hear his own modest proposals about the former extent of local glaciers transformed into a cataclysmic vision of a world dominated by ice. Alexander von Humboldt, one of Europe's leading scientists and a generous mentor of Agassiz, counseled his protégé to abandon "these general considerations (a little icy besides) on the revolutions of the primitive world—considerations which, as you well know, convince only those who give them birth."

But Agassiz would not renounce his heresy. Instead, he turned his energies to proselytizing. One of his main targets was the Reverend William Buckland, Oxford University's first professor of geology, who had made the dramatic discovery of fossil remains at the Kirkdale cave in 1821. Buckland—who attributed the Kirkdale fossils to the action of the Flood—was regarded by many as the finest geologist of his time. He was so enthusiastic in pursuit of his discipline that his horses were said to stop automatically at every rock outcrop; and he was so knowledgeable about the geology of England that once, when he and some companions lost their way riding back to London on a dark night, he had only to dismount, scoop up a handful of earth and smell it to pronounce that they were near Uxbridge. Agassiz's problem was that Buckland remained an ardent believer in the Flood. "The grand fact of a universal deluge at no very remote period," he had written in 1819, "is proved on grounds so decisive and incontrovertible, that, had we never heard of such an event from Scripture, Geology, of itself, must have called in the assistance of some such catastrophe to explain the phenomena of diluvial action."

Agassiz had met Buckland in 1834, when the young Swiss scientist visited England to examine fossil-fish collections. The two became friends, and in 1838 Agassiz got the chance to expound his ice age views when Buckland attended a meeting in Freiburg, Germany. Although the professor's faith in the Flood remained unshaken, he later accompanied Agassiz into the Swiss mountains to examine the evidence that had convinced the younger man. What Buckland saw caused him to abandon some of his cherished notions about the Flood, but back in England he was assailed by doubts: When his wife wrote to Agassiz thanking him for his hospitality, she confessed that her husband was "as far as ever from agreeing with you."

In August 1840, Agassiz received from Alexander von Humboldt a letter indicating that the eminent German's opposition might be weakening. "I cannot close this letter," Humboldt wrote, "without asking your pardon for some expressions—too sharp, perhaps—in my former letters about your geological conceptions." Two months later, Agassiz published an expanded version of his "Discourse of Neuchâtel" in a book called *Studies on Glaciers*, but this work engendered even more antagonism than his original speech had aroused. The hostile reception was due as much to professional jealousy as to honest scientific disagreement. De Charpentier, who was planning a book of his own, was furious that Agassiz had beaten him into print; Karl Schimper, originator of the term Ice Age, was outraged because he felt that his ideas had been insufficiently acknowledged in Agassiz's book, and its publication opened a breach between them that was never closed.

Also in 1840, Agassiz visited Scotland and addressed the annual meeting of the British Association for the Advancement of Science. This prestigious event attracted not only Buckland but also such prominent British geologists as Charles Lyell and Roderick Impey Murchison, president of the Geological Society of London. Once again Agassiz presented his theory—more sweeping than ever, for by now he was convinced that the ice sheets had once covered all of the northern parts of Europe, America and Asia. Again his arguments fell mainly on disbelieving ears. Lyell was particularly explicit in his denunciation; Buckland, however, remained silent throughout.

It was a thoughtful silence; Buckland had been finding it increasingly difficult to explain some landforms as the consequences of a universal deluge, and he was beginning to reconsider the possibility of glacial action. After the meeting, the professor invited Agassiz and Murchison to join him on a field trip to northern England, Scotland and Ireland. For Buckland, the journey was a revelation. During his earlier visit to Switzerland, he had been briefly swayed by the evidence of ancient glacial action there, but had persuaded himself that it was purely local in nature; now in Scotland, a country where no glaciers existed, Agassiz pointed out to him the same kinds of evidence—moraines, erratics, and grooved and polished rocks. Approaching a valley, Agassiz told his companions exactly where they would find a terminal moraine, and he was right. In Glen Roy, Agassiz convinced Buckland that the famous "parallel roads" of the valley—three stepped terraces that resemble roadways cut into the slopes—are in fact the successive shorelines of an ancient lake that had been dammed and much enlarged by a glacier.

By the end of the trip, Buckland had moved firmly into the ice age camp, and he lost no time in winning over another important convert—Charles Lyell. Soon after Agassiz had returned home, he received an enthusiastic letter from Buckland stating that "Lyell has adopted your theory *in toto!!!*" Murchison, however, remained unimpressed, and the strength of his opposition to the glacial theory delayed its acceptance by the rest of the British scientific world.

Later in 1840, Agassiz was invited to address the Geological Society of London. Buckland, vice president of the society, and Charles Lyell both

presented papers defending Agassiz's case, but most members of the society were highly skeptical. "Does Professor Agassiz suppose that Lake Geneva was occupied by a glacier 3,000 feet thick?" demanded one listener. "At least," replied Agassiz. This response stung his questioner into remarking that the glacial theory was "the climax of absurdity in geologicial opinions." Murchison gloomily observed that there would be no stopping the glaciologists if their theories were accepted; next, he said, they would be saying that glaciers had once overrun Hyde Park—which, in fact, they never had. But the newly converted Buckland concluded the meeting by cheerfully condemning to "the pains of eternal itch without the privilege of scratching" anyone who dared to challenge the evidence that supported the glacial theory.

Even so, when Murchison delivered his annual presidential address to the Geological Society of London two years later, he devoted the last part of his lengthy talk to a spirited attack on the ice age theory. There was "little risk that such doctrine should take too deep a hold of the mind" in Europe, Murchison assured his audience, but he was concerned that Agassiz's views might be championed by people of a more susceptible nature. Already, he told his listeners, he had read disturbing news that an American geologist, Edward Hitchcock, had recently stated, in Murchison's phrase, that "the work of Agassiz unexpectedly threw a flood of light upon his mind," enabling him to explain such phenomena as gravel deposits and striated rocks. (Another American, Timothy Conrad, had declared his support for Agassiz only two years after the "Discourse of Neuchâtel"; in 1839, Conrad published a paper claiming that the rocks of western New York State had been polished and striated by glaciers in the same way that the rocks of the Jura had been.) The truth was that the seeds planted by Agassiz and nurtured by Buckland and Lyell had taken root in the New World as well as the Old.

Opposition from diehard believers in the Flood doctrine would continue throughout the 19th Century, and one British geologist went so far as to claim that polished rocks in Wales had been rubbed smooth by small boys sliding down them. Murchison—splendidly defiant to the last—declared in 1864 that "ice, *per se,* neither has nor has had any excavating power." But the last effective resistance to the glacial theory withered in the mid-1860s, after the Scottish geologist Thomas Jamieson published a persuasive paper that compared the observed effects of flooding and glaciation, and showed that only glacial action could account for the erratic boulders and striated bedrock found in Scotland. Archibald Geikie, another prominent Scottish geologist, was so thoroughly convinced by Jamieson's arguments that he wrote: "Doubtless, the glacial theory will ere long come to be universally accepted in this country as it ought to have fully twenty years ago, when its first outlines were sketched by Agassiz."

By then, the author of the theory was living in the United States; he had moved there in 1846 and had subsequently been appointed Professor of Zoology at Harvard. In his adopted country, Agassiz increasingly turned his attention to his first love, the study of fossil fish, and he died in 1873 without contributing anything more of major significance to the glacial debate. But he had opened the eyes of his fellow scientists to the Ice Age world, and he lived long enough to see his theory vindicated. He left it to others to explore that world, give it a chronology, and attempt to solve the most intriguing question of all: What had caused the Ice Age?

A bed of limestone along the coast of St. John's Bay, Newfoundland, displays evidence of an ice sheet's inexorable movements; the surface was grooved and polished by the abrasive debris trapped beneath millions of tons of ice.

Even as the controversy raged, some geologists were pondering the implications of massive glaciation. As far back as 1832, in fact, Charles Lyell had proposed that changes in sea level were caused by the fluctuations in the amount of land ice; and in 1841, after the publication of Agassiz's theory, a Scottish geologist named Charles Maclaren pointed out that if glaciers had indeed once covered the summits of the Jura, "the abstraction of the water necessary to form the said coat of ice would depress the ocean about 800 feet." Maclaren had no way of knowing the maximum volume of the ice sheets, but by the 1870s, geologists in Europe and North America had determined the ancient boundaries of the ice by charting the locations of glacial deposits, and had calculated the average thickness of the ice sheets by examining the vertical extent of their effects on mountains. From such measurements, an American geologist, Charles Whittlesey, concluded in 1868 that "at the period of greatest cold, the depression of the ocean level should be at least three hundred and fifty or four hundred feet."

Such estimates were apparently contradicted by the finding of marine fossils far above modern sea levels, indicating that some shorelines had in fact been higher immediately after the Ice Age than they are at present. But Thomas Jamieson, who had been much intrigued by the discovery of marine deposits in Scandinavia more than 1,000 feet above sea level, offered a brilliant explanation for the paradox in 1865. "In Scandinavia and North America, as well as in Scotland," he wrote, "we have evidence of a depression of the land following close upon the presence of the great ice-covering; and, singular to say, the height to which marine fossils have been found in all these countries is very nearly the same. It has occurred to me that the enormous weight of ice thrown upon the land may have had something to do with this depression." Anticipating the conclusions of 20th Century geophysicists, Jamieson explained the glacial depression and the postglacial uplift by postulating that beneath the earth's rigid crust is a layer of rock, "in a state of fusion," that would yield under the pressure exerted by a massive ice sheet and gradually rebound as the covering melted.

Clues to the climatic conditions that had prevailed during the Ice Age were found in different parts of the globe. One key discovery was that deep beds of loess—fine, windborne sediments deposited during periods of glaciation—are similar to more recent deposits transported from the cold, dry Gobi Desert. This was one of the first indications that the glacial environment had been arid. In the American Southwest, however, geologists found evidence that the region had been much wetter than it is today; during the 1870s, Grove Karl Gilbert of the U.S. Geological Survey confirmed from his examination of fossil shorelines that Lake Bonneville had covered 20,000 square miles. Today, the only remnant of this huge prehistoric body of water is the Great Salt Lake of Utah.

As such studies accumulated, geologists realized that the periodic advances and retreats of glaciers were related to past warming and cooling trends in global climate. Louis Agassiz had supposed that the Ice Age was a unique episode in the history of the earth; but in the 1840s, the English geologist Joshua Trimmer found two distinct layers of till—sediments deposited by a glacier—in a cliff on the coast of East Anglia, which suggested that there had been at least two separate glaciations in Britain. In the 1860s, Archibald Geikie argued that a layer of plant fragments found between two layers of Scottish till represented a period of warm climate be-

After the Ice, the Deluge

In the 1960s, after more than a century during which scientists had prowled the terrain of every continent searching for evidence of ancient glaciation, they tapped an entirely new source of information—an Ice Age landscape that has been hidden beneath the waves since the massive ice sheets began to retreat some 18,000 years ago.

The great glaciers that once invaded the Northern Hemisphere were so enormous—almost two miles thick in places—that when they melted, they caused one of the greatest floods in geologic history; eventually, the sea level rose some 400 feet worldwide. For the most part, the water level came up slowly—about two feet per century—but even this gradual rise was enough to inundate low-lying coastal areas at a rapid pace. The western coastline of the Gulf of Mexico, for example, was so flat that the sea overran dry land at a rate up to 200 feet per year.

Throughout the world, fertile grasslands became first swamps and then sea floors. Coastal forests were submerged so quickly that instead of decomposing, the doomed trees were preserved, virtually pickled in the briny water. In the tropics, many hills became islands and were soon ringed by colonies of coral, which thrive in warm shallows. As the centuries passed, many of the islands were submerged by the seas while the encircling coral reefs continued to grow upward toward the surface. Eventually, only ringlike atolls teeming with fish remained as living halos commemorating the antediluvian hilltops.

Working more than 70 feet beneath the water's surface, a diver takes measurements near an ancient tree stump, a remnant of a forest of willows and alders that flourished on the coast of Japan some 10,000 years ago.

Coral reefs some 300 miles off the coast of India form atolls around islands that were diminished or engulfed by the rising sea. Coral can grow several inches a year to stay in its optimum environment (within 20 feet of the surface); the reefs thus rose at about the same pace as the flood of water that was released by the melting glaciers of the Ice Age.

The stalactites and stalagmites bedecking
this submarine cave in the Bahamas are proof
that the level of the sea here was at one time
much lower; such formations grow only in air.

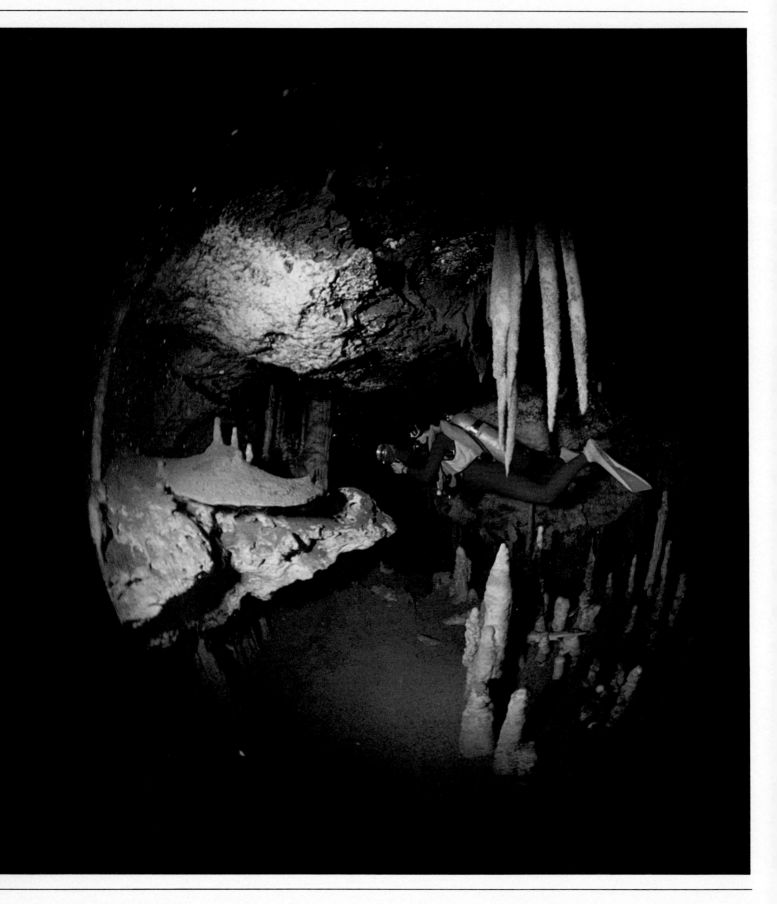

tween two advances of the ice sheet; the argument was settled conclusively in the 1870s when the remains of a forest were found sandwiched between two sheets of till in the American Midwest.

With the discovery that the Ice Age had not been a single period of glaciation, scientists set out to establish a sequence and chronology for the succession of glacial advances and withdrawals. To gauge the date of events that had taken place long before recorded history, geologists applied Hutton's law of uniformitarianism: First they measured the speed of geological processes that occur at what they assumed was a more or less constant rate, then they calculated how long it had taken to shape the features formed by these processes. All that was needed was a starting point—and a little ingenuity.

In the late 19th Century, for example, Grove Karl Gilbert followed up on the work of earlier geologists and sought to calculate the time that had elapsed since the last glaciation by measuring the distance that Niagara Falls had receded up its gorge since the retreat of the ice. A deposit of glacial debris some five and a half miles below the falls gave him his starting point, and historical records revealed the average rate at which the lip of the falls had been cut back by erosion. By measuring the distance between the glacial deposit and the falls, and then dividing this distance by the presumed annual rate of erosion, Gilbert estimated that the glacier had retreated from the gorge about 7,000 years ago—a date far more recent than previous estimates of 30,000 years and more.

As a means of establishing a geological calendar, the dating techniques used by Gilbert and others were crude, since they could not accurately take into account the variations that affect all natural processes; Gilbert, for example, had no way of establishing whether the rate at which the Niagara gorge eroded had changed over the millennia because of alterations in the flow of the river or differences in the hardness of the rock. What was needed was a method based on a process with an annual rhythm that left a clearly readable imprint.

Such a method had, in fact, been employed in 1878 by Baron Gerard de Geer, a Swedish geologist. During field work in the Stockholm region, de Geer was struck by the regularity of the laminations in the sediments at the bottoms of lakes fed by glaciers. When glacier ice melts during the summer, de Geer discovered, the meltwater carries off a load of debris that settles in the nearby lake to form a distinct profile for that particular year. The heavy material sinks first, forming a coarse layer, while the lighter material remains longer in suspension, eventually accumulating as a fine sediment on top of the coarse deposits.

Since glacial lakes are born as soon as the glaciers retreat, de Geer was able to estimate the age of each lake—and thus the approximate date of the glacial recession from that locality—simply by counting the pairs of sediments, called varves, on its bed. In addition, because varves vary in thickness according to the climate, being thickest in warm years when melting glaciers release large quantities of sediment, de Geer was able to plot the progress of the Scandinavian Ice Sheet's retreat by correlating the sediment patterns of different lakes. He calculated that the oldest lakes—those nearest the margin of the ancient ice sheet—had been formed approximately 12,000 years ago, while the youngest were little more than 6,000 years old. And the fluctuations in the thickness of the varves gave

him a crude picture of the climatic record since the retreat of the glaciers.

Such natural calendars cover only the postglacial period; different dating methods are needed to calculate the duration of ice ages and interglacials that make up an ice epoch lasting for millions of years. But even establishing the succession of glacials and interglacials was difficult, since advancing ice sheets act like bulldozers, all but destroying previous deposits. By the end of the 19th Century, however, geologists had overcome these problems and uncovered evidence of four separate glacial tills in North America, which strongly suggested that the continent had experienced at least four ice ages in recent geologic time. In order of occurrence, these ice ages were named after the states where the evidence of their effects had been most studied: Nebraskan, Kansan, Illinoian and Wisconsinan.

Using different methods for identifying glacial chronologies, two German geographers, Albrecht Penck and Eduard Brückner, found that the Alpine region of Europe had also been affected by four ice ages, which they named—from oldest to youngest and in alphabetical order—after four of the river valleys where they had collected their evidence: Günz, Mindel, Riss and Würm. By analyzing varves from Swiss lakes, Penck and Brückner estimated that ice sheets withdrew from the Alps about 20,000 years ago; then, by comparing the depth of postglacial erosion with the measurement of erosion that had occurred during each of the interglacials, they concluded that the interglacial before the most recent ice age lasted 60,000 years, that the preceding interglacial lasted about 240,000 years, and that the entire ice epoch had spanned a period of 650,000 years. Their evidence—eventually published in 1909—was impressive, and most scientists believed that the problem of establishing the chronology of the ice ages had been solved.

While some investigators were laboring to map the succession of ice ages and assign them to slots on a geological timetable, others sought to determine what had caused the ice sheets to descend from their polar and mountain preserves and then to withdraw or vanish. It was, to say the least, a daunting enterprise. In 1884, the American geologist Clarence Edward Dutton explained the difficulties facing anyone attempting to solve the mystery. "While it cannot be doubted that the climate of the glacial period was in some important respects different from the present climates," he wrote, "the moment any attempt is made to ascertain what would be the effect if any one of the determinants of climate were to undergo a marked variation, it is found to be so intricately interwoven with many other conditions and determinants that the investigator is generally baffled in his endeavours to assign a just and proper weight to them all."

In effect, the earth's atmosphere and oceans make up a complex system that is powered largely by heat from the sun. But much of the sun's energy never finds its way into the system; about 50 per cent of it is reflected back into space by clouds and by the atmosphere itself. And of the radiation that does reach the planet, something like a third is reflected by various surfaces. New snow, for example, reflects 90 per cent of the sunshine that strikes it, while sea ice and deserts reflect about 35 per cent. On the other hand, land areas such as forests reflect as little as 10 per cent of the solar radiation that hits them, and oceans sometimes reflect a scant 3 per cent.

The solar radiation that is finally absorbed by the earth is not evenly distributed. The equatorial regions absorb more radiation than they reflect,

Desiccated salt flats and terraced shorelines commemorate Utah's Lake Bonneville—a prehistoric inland sea that once covered 20,000 square miles. When the last ice age ended about 10,000 years ago, the region's climate became arid and the huge lake evaporated.

The underlying rock edge of Canada's Horseshoe Falls is cut back some one to three feet each year by the rushing Niagara River. Attempts to use the

erosion rate to date the end of the last ice age, when the cascade was formed, have been unsuccessful since, over time, the erosion rate has varied widely.

while the polar regions reflect more than they absorb; only at lat. 40° N. and lat. 40° S. are absorption and reflection approximately balanced. If the differences in the local heat balances were not evened out, the Equator would get steadily warmer and the Poles ever colder. But the earth's oceans and wind currents transfer heat from the tropics toward the Poles while carrying cooling water and frigid air from the Poles toward the Equator.

Any significant cooling of one part of the earth's climatic system will bring about corresponding changes in the other parts of the global network, perhaps triggering a set of reactions that could bring about an ice age. The problem lies in identifying the initial change and its cause. To complicate matters further, any valid theory of the cause of ice ages has to explain what process brings them to an end and why they have occurred so frequently.

In the years after Louis Agassiz first presented his ice age theory, scientists who accepted the idea came up with a number of explanations for widespread glaciation. Some thought that the ice sheets were brought on simply as a result of the earth's cooling from a previously hot state; others fancied that the entire solar system was passing through alternate cold and warm regions of space. But the most widely held notion was the one championed by Charles Lyell, who maintained that glaciation—and subsequent warming—must have resulted from presumed vertical movements of the earth's crust, continuous risings and fallings that turned lands into seas and seas into lands. According to Lyell, extensive areas of land surfacing at high latitudes would lead to increased snowfall and thus encourage the formation and growth of ice sheets; during periods when large land areas were exposed along the Equator and in the tropics, Lyell theorized, these lands would become heated by the sun and generate warm winds that would ameliorate climates in the upper latitudes.

If Lyell's explanation seemed plausible enough to most scientists, it did not satisfy the Scottish geologist James Geikie, brother of Archibald Geikie. In 1874, writing in the first edition of his classic text, *The Great Ice Age,* Geikie roundly dismissed the Lyell theory, pointing out that it ignored, among other things, the warming influence of ocean currents. But Geikie was not so reckless as to demolish a popular theory of the causes of ice ages without having in mind a plausible substitute. He had embraced an ingenious explanation that was nearly as old as the ice age theory itself.

In 1842, only five years after Louis Agassiz delivered his "Discourse of Neuchâtel" to a skeptical audience, a French mathematician named Joseph Alphonse Adhémar applied his broad knowledge of astronomy to the ice age problem. His conclusion: The earth's climate is influenced by its orbital path and by variations in the angle of its axis in relation to the sun. The fact that the earth's physical relationship to the sun undergoes constant and significant change is evident from the changing seasons. If the earth were positioned so that its axis of rotation formed a right angle to the rays of the sun, there would be no seasons. But the earth's axis is tilted by about 23 degrees, and as the planet makes its annual circuit of the sun, each Pole points first toward the sun and then away from it. The North Pole is inclined most toward the sun on June 21, making that day the longest of the year and marking the start of summer in the Northern Hemisphere; the Arctic regions are tilted farthest from the sun on December 21, the shortest day of the year and the hemisphere's first day of winter. In the Southern

Hemisphere, the seasons are reversed. Twice each year, on March 20 (the first day of spring in the North) and September 22 (the first day of fall), the North and South Poles are the same distance from the sun; on these dates, known as equinoxes, the hours of darkness equal the hours of daylight all over the world.

The angle of the earth's axial tilt is not constant. In fact, over a period of some 41,000 years, it varies between about 22 degrees and more than 24 degrees (at present, the angle is about 23½ degrees and getting smaller). As it decreases, the differences between seasons tend to diminish, with winters becoming milder and summers cooler. When the angle is greatest, the contrast between the seasons is most marked. The effect of this change is the same in both hemispheres.

Further complicating matters, the earth does not maintain a constant distance from the sun. As the astronomer Johannes Kepler first established in the 17th Century, the planet's orbit around the sun is not a circle but an ellipse—which means that at some times in the year the earth is closer to the sun, at other times farther away. At present, the earth reaches its nearest point to the sun, or perihelion, in January, and reaches aphelion—the point farthest from the sun—in July. The difference between the earth's distance from the sun at aphelion and its distance at perihelion is about three million miles. When the earth approaches perihelion, it accelerates in its orbit; when it nears aphelion, it slows down. The result is that the cool seasons in the Northern Hemisphere are shorter than the warm seasons by about seven days, while in the Southern Hemisphere, the warm seasons are seven days shorter than the cool ones.

Adhémar also knew that the earth's geometric relationship to the sun is affected by a third process: a slow gyration, or wobble, of the earth's axis of rotation. Since the earth itself is not round but oval—with a slight bulge at the Equator—the gravitational pull of the other bodies in the solar system causes it to wobble, just as the earth's gravity makes a slowly spinning top wobble. This effect was first identified by the Greek astronomer Hipparchus in about 120 B.C., when he compared his own astronomical measurements with those made 150 years earlier by Timocharis. He found that the positions of the stars differed: Because the earth's axis wobbles, it does not always point toward the same stars. At present, the North Star—so called because the North Pole points toward it—is Polaris; it remains fixed while the other stars seem to rotate. But 4,000 years ago, the polestar was Thuban in the constellation Draco; and in 12,000 years, the North Pole will point to another star, Vega. In 22,000 years—the length of time it takes the axis to complete one wobble—Polaris will again be the polestar.

The effect is more scientifically known as the precession of the equinoxes, because in addition to altering the map of the universe as seen from the earth, it affects the length of the seasons by changing the time of the year during which the earth is nearest and farthest from the sun. Northern Hemisphere winters are shorter and warmer now than winters in the Southern Hemisphere because the perihelion comes in January. Eleven thousand years ago, however, the slow wobble of the precession cycle had tipped the earth so that January occurred during aphelion, when the planet is at its greatest distance from the sun. Thus winters in the Northern Hemisphere were then longer than winters in the Southern Hemisphere.

Adhémar attempted to convince his colleagues that this cycle was the cause of ice ages. He suggested that whichever hemisphere had a longer winter would experience an ice age, so that every 11,000 years glacial conditions would occur in either the Northern or Southern Hemisphere. But he had ignored one important point, which was noted by the German naturalist Alexander von Humboldt in 1852. Any decrease in the solar radiation received by a hemisphere while it is tilted away from the sun, Humboldt observed, is balanced by a corresponding increase in solar radiation during the opposite season, when the earth's position has changed so that the tilt is toward the sun. In other words, the hemispheres receive roughly equal amounts of solar energy in the course of a year, and the fact that one has a longer winter than the other could not in itself precipitate an ice age.

Humboldt had shown that Adhémar was wrong, but he had not destroyed the notion that there is a connection between ice ages and the earth's relationship to the sun. The theory would soon be revived and refined, though the man who picked up the threads of Adhémar's argument was a most unlikely pioneer of science.

James Croll, the son of a stonemason, was born in Perthshire, Scotland, in 1821, and received only the rudiments of a formal education before leaving school at the age of 13. When he was only 11, however, an incident occurred that, according to Croll's autobiography, "led to a new epoch in my life." He bought the first issue of *Penny Magazine,* a wide-ranging and entertaining publication for children. Young Croll was so fired with enthusiasm by its contents that he became an avid reader of nearly anything he could get his hands on. Shortly afterward, Croll recalled, he procured his first books on physical science. "At first I became bewildered," he wrote,

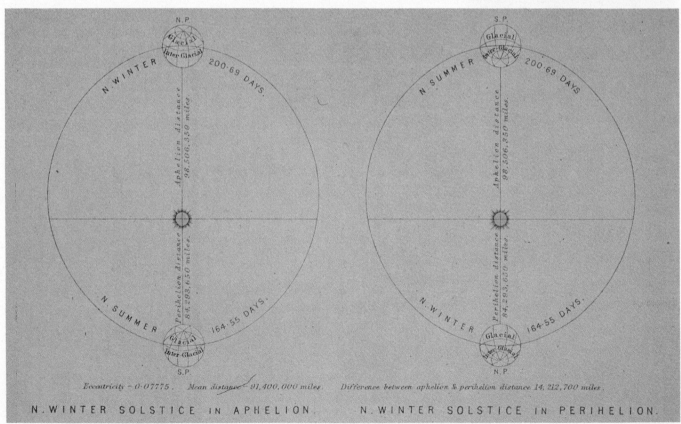

N. WINTER SOLSTICE IN APHELION.   N. WINTER SOLSTICE IN PERIHELION.

"but soon the beauty and simplicity of the conceptions filled me with delight and astonishment, and I began then in earnest to study the matter."

Concentrating on the laws and principles rather than experimental details, Croll managed to gain a "pretty tolerable knowledge" of mechanics, pneumatics, hydrostatics, and the nature of light, heat, electricity and magnetism by the time he was 16 years old. Ironically, while the results of his studies would in time be used to explain the findings of geology, he had no interest in that branch of science; it "appeared so full of details and so deficient in rational principles," he noted. At the end of his life, Croll would claim that "geology is almost the only science on which I have never spent a day's earnest study."

The self-taught scholar's extraordinary efforts to master the principles of science were curtailed when Croll was 16. Required to earn a living, he became apprenticed to a millwright. For five years he led a nomadic existence, traveling from farm to farm to repair machinery, often sleeping in barns where he and his fellow apprentices "had to bury ourselves under the clothes to secure protection from the rats." Tiring of such a life, Croll turned to carpentry, but was forced to abandon this trade when he aggravated an old arm injury. All his life, in fact, Croll was to be plagued by illnesses, some real and some perhaps imaginary.

Failure became a habit for him. He opened a teashop, but shyness and a recurrence of the arm injury kept him from attending properly to his customers and forced him to close the business. He tried to run a temperance hotel in a small Scottish town that had a total of 16 inns and pubs; the venture met a quick end. He then tried to sell life insurance, an enterprise that he described as "four and a half years of about the most disagreeable part of my life."

Actually, Croll seems to have brought little enthusiasm to any of these enterprises, and may even have been secretly relieved when they failed and left him with more time to spend on his scholarly pursuits. During a period of unemployment after his distasteful career as an insurance agent, Croll published a small volume called *The Philosophy of Theism,* which actually enjoyed a small success. An interlude as a writer on a Glasgow newspaper devoted to the cause of temperance followed, and then Croll got the most pleasant job he ever held: He became a janitor at the Andersonian College in Glasgow.

"Taking it all in all," Croll wrote later, "I have never been in any place so congenial to me as that institution proved. After upwards of twenty years of an unsettled life, full of hardships and difficulties, it was a relief to get settled down in what might be regarded as a permanent home. My salary was small, it is true, little more than sufficient to enable us to subsist; but this was compensated by advantages for me of another kind. Here was the fine scientific library, belonging to the Glasgow Philosophical Society, to which I had access—a privilege of which I took advantage. My duties were regular and steady, requiring little mental labour; and as my brother was staying with me, he gave me a great deal of assistance, which consequently allowed me a good deal of spare time for study."

Among the subjects studied by Croll was the Ice Age. "At this period," he explained, "the question of the cause of the Glacial epoch was being discussed with interest among geologists. In the spring of 1864 I turned my attention to this subject." Naturally enough, Croll's studies led him to

Scotsman James Croll (*top left*) stunned the scientific world in the 19th Century by showing how ice ages might be caused by variations in the orbit of the earth. Periodically, he said, the elliptical orbit becomes more elongated, and for about 10,000 years one hemisphere's winter occurs at aphelion—the orbit's farthest point from the sun (*far left*). When this takes place, Croll proposed, an ice age ensues in that hemisphere. Conversely, when winters coincide with perihelion—the point nearest the sun (*left*)—a warm interglacial period results.

Joseph Adhémar's discredited claim that severe climate changes could be brought about by the precession of the equinoxes, the wobbling of the earth's axis. But Croll now had a distinct advantage over Adhémar, for he also knew of the later work of the French astonomer Urbain Jean Joseph Leverrier, who had discovered that the shape of the earth's path around the sun changes over time. Leverrier's calculations had shown that during a period of some 100,000 years, the eccentricity of the earth's orbit—sometimes called stretch—varies considerably, from a nearly perfect circle to a pronounced ellipse. And the more eccentric the orbit, the farther the earth's circuit takes the planet from the sun.

Croll decided that ice ages were caused by these orbital changes. In August 1864, he published a paper on the subject in the *Philosophical Magazine,* a twice-yearly science journal. According to Croll, the paper "excited a considerable amount of attention, and I was repeatedly advised to go more fully into the subject; and as the path appeared to me a new and interesting one, I resolved to follow it out. But little did I suspect, at the time when I made this resolution, that it would become a path so entangled that fully twenty years would elapse before I could get out of it."

Using Leverrier's calculations, Croll proceeded to plot a curve showing changes in the orbit's shape during the past three million years, and deduced that ice ages occurred during periods when the earth's orbit was highly elliptical. Since he accepted the fact that variations in the earth's orbit do not affect the total amount of radiation received during a year, Croll concluded that it must be the seasonal effect of the orbital eccentricity that produced ice ages. He reasoned—incorrectly, as it turned out—that less sunlight in winter encouraged the accumulation of snow, which would then reinforce the seasonal cooling by reflecting incoming radiation. From there he went on to determine that the precession of the equinoxes—wobble—and the degree of orbital eccentricity were the important factors governing the amount of winter sunlight received by the earth.

If a hemisphere's winters occur when it is inclined away from the sun during the period of earth's greatest orbital eccentricity, Croll concluded, they would be longer by as much as 36 days, and colder than usual, possibly for thousands of years—cold enough and long enough to trigger an ice age. And since the two hemispheres take turns experiencing seasons at aphelion, it followed that ice ages would be likely to occur in the Northern Hemisphere during an 11,000-year span, or half of the precessional cycle, and then in the Southern Hemisphere during the next 11,000 years. For the balance of the 100,000-year orbital cycle, the earth's orbit would be less eccentric, and there would be no ice ages.

Croll's continuing papers on the astronomical theory were greeted with enthusiasm by scientists who had accepted the idea of ice ages but who had yet to find a satisfactory explanation for such major climate changes. Archibald Geikie was so impressed by his fellow Scot's work that he offered him a post at the Geological Survey of Scotland, of which Geikie was director. In 1867, Croll accepted the job and moved to Edinburgh to continue his research. In 1875, he published a book outlining his latest findings, which took into account the effect of the earth's tilt. Although the length of this cycle was not yet known, Croll hypothesized that an ice age would be most likely to occur when the earth's angle of tilt is least, for then the polar regions receive the smallest amount of heat.

Following the appearance of his book, which he titled *Climate and Time,* Croll was showered with academic honors. He was elected a Fellow of the Royal Society in London and chosen as an honorary member of the New York Academy of Science. The University of St. Andrews presented him with an honorary doctoral degree, and the Geological Society of London granted him funds to aid his research. Croll's theory had taken firm hold in the scientific world, and the eminent James Geikie was moved to write: "The astronomical theory would appear to offer the best solution to the glacial puzzle. It accounts for all the leading facts, for the occurrence of alternating cold and warm epochs, and for the peculiar character of glacial and interglacial climates."

Then a major problem began to emerge. According to Croll's calculations, the last ice age had peaked about 80,000 years ago, but after the publication of his theory, studies of the geological record in both Europe and North America made it clear that glacial conditions had persisted until a much more recent date. By the time of Croll's death in 1890, most geologists were beginning to believe that his theory must be wrong. Even James Geikie, while insisting that Croll's work had "undoubtedly thrown a flood of light on our difficulties," expressed misgivings about the astronomical theory. But Geikie remained hopeful: "It may be," he added, "that some modification of his views will eventually clear up the mystery. But for the present we must be content to work and wait."

Milutin Milankovitch, the man who would provide the modification that Geikie hoped for, was born in the politically volatile Balkan nation of Serbia—later incorporated into Yugoslavia—in 1879. Unlike Croll, Milankovitch came from a relatively privileged background. His family owned extensive farmlands and vineyards, and several of his relatives were university graduates. For a time, he gave in to family pressure and studied agriculture in preparation for taking up the management of the Milankovitch estates. But he was more interested in the sciences and went instead to Vienna, where he earned a doctorate in engineering in 1904. After working for five years as an engineer in Vienna, building such things as dams and bridges of reinforced concrete, he gladly returned to his homeland to accept a post at the University of Belgrade. There he lectured on mechanics, astronomy and theoretical physics, and yearned for a challenge that would permit him to make his mark in the world of science.

In 1911, during an evening of wine tippling with a poet friend, Milankovitch selected his challenge: He would develop a mathematical theory that would enable him to determine not only the temperature of the earth at different latitudes and at different times but also the climates of other planets in the solar system. It was to be an ambitious scientific sojourn in what Milankovitch called "distant worlds and times," and the young professor had picked an ideal stage of his life to begin. "I set out on this hunt in my best years," he recalled later. "Had I been somewhat younger I would not have possessed the necessary knowledge and experience. Had I been older I would not have had enough of that self-confidence that only youth can offer in the form of rashness."

Milankovitch pursued his goal with single-minded devotion. As he later observed: "When a scholar stands before a scientific problem, he becomes like a hunting dog that has sensed the game." His first step was to make a

An Interplay of Celestial Effects

The first step toward the astronomical theory of the ice ages was taken 2,000 years ago, when the Greek astronomer Hipparchus discovered that the earth wobbles like a top as it whirls through space. By the end of the 19th Century, astronomers had described in detail three critical ways in which the earth's position in space and its orbit around the sun change in response to the gravitational tugs of sun, moon and planets.

The first, Hipparchus' discovery, is called axial precession. The second is tilt—an ever-changing angle between the earth's axis and the plane of the planet's orbit around the sun. The third factor is the eccentricity, or varying shape, of the orbit itself, which ranges from nearly circular to a pronounced ellipse. Between 1912 and 1941, mathematician Milutin Milankovitch performed exhaustive calculations to show how all of these factors—precession, tilt and eccentricity—could cause changes in the intensity of summer sunshine extreme enough to explain the recurrence of ice ages. Extensive geological evidence discovered since Milankovitch's time confirms the impact of these celestial cycles on the earth's climate.

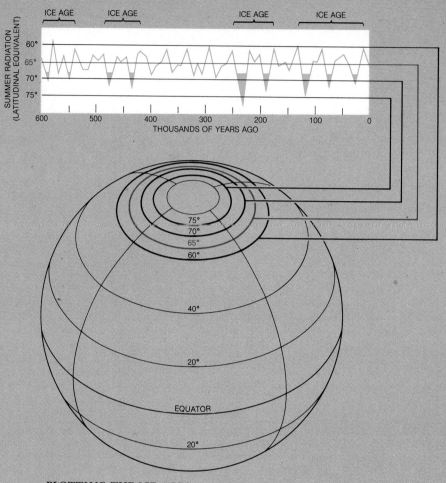

PLOTTING THE ICE AGES
A graph plotted by Milankovitch compares the summer solar radiation that has reached lat. 65° N. during the past 600,000 years with present-day summer radiation at other latitudes. The shaded dips coincide with periods when Europe was experiencing glaciation; 25,000 years ago, summer solar radiation at lat. 65° N. equaled the amount received today more than 400 miles north, at lat. 71° N.

THE 100,000-YEAR STRETCH
The orbit of the earth gradually stretches from nearly circular to an elliptical shape and back again in a cycle of approximately 100,000 years. During the cycle, the distance between earth and sun varies by as much as 11.35 million miles.

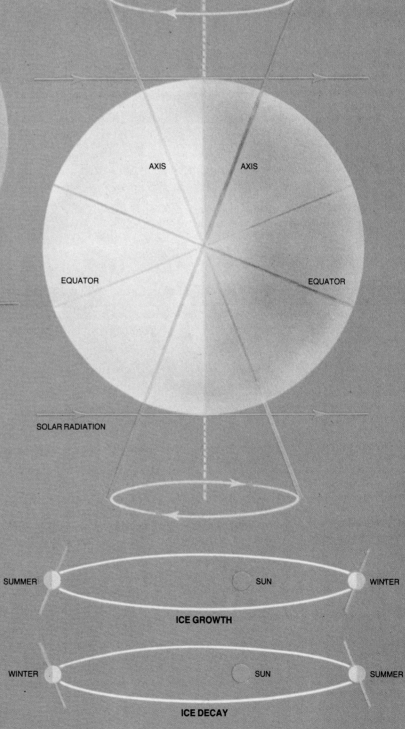

THE 41,000-YEAR TILT

The earth's axis is never perpendicular to the plane of its orbit; over the course of about 41,000 years the angle varies between 21.5 and 24.5 degrees. Because of the tilt, the solar radiation striking any point on earth fluctuates during the yearly orbit, producing seasons. When the tilt is greatest, summers are hottest, winters are coldest.

THE 22,000-YEAR WOBBLE

Even while the shape of its orbit and the tilt of its axis are changing, the earth wobbles slowly in space, its axis describing a circle once every 22,000 years *(top right)*. Because of this movement, known as axial precession, the distance between the earth and the sun in a given season slowly changes. Today, for instance, the shape of the orbit places the planet closest to the sun in the Northern Hemisphere's winter and farthest away in summer. The combination *(right)* tends to make winters mild and summers cool — and favors ice-sheet growth. However, 11,000 years ago, the arrangement was just the opposite *(bottom right)*, setting the stage for the Northern Hemisphere ice sheets to decay.

thorough survey of work that had already been done in his chosen field. He was fascinated by James Croll's astronomical theory, but concluded that Croll, for all his considerable accomplishments, had lacked the precise data required to deal adequately with a problem of such magnitude. Luckily, though, Milankovitch came across the more recent studies of the German mathematician Ludwig Pilgrim, who in 1904 had published minutely detailed calculations of the precession of the equinoxes and changes in the earth's orbital eccentricity and angle of tilt. Indeed, Pilgrim had even gone so far as to chart the relationship between orbital eccentricity and the presumed chronology of past ice ages. Milankovitch judged that Pilgrim's understanding of climatology left much to be desired, but could find no fault with his mathematics; he used the German's figures to work out his own calculations of past climates of the earth and other planets.

His progress was interrupted in the fall of 1912 by the outbreak of the First Balkan War, in which Serbia joined its neighboring allies to expel the Turks from southeastern Europe; a reserve Army officer, Milankovitch was called to active duty with his regiment. Hostilities were short-lived, however, and Milankovitch soon returned to his civilian desk. During the next two years he published several papers outlining the emerging results of his work, which indicated that glacial advances and retreats could indeed be brought about by changes in solar radiation due to the precession of the equinoxes and to variations in the earth's orbital eccentricity. (His calculations became considerably more accurate after 1913, when American scientists at the Smithsonian Institution were able to establish the solar constant, or the intensity of the sun's radiation.) He also showed that variations in the planet's angle of tilt influenced climate to a far greater degree than James Croll had believed.

In the summer of 1914, war intruded once again on Milankovitch's affairs. He was visiting his home village of Dalj—then a part of Austria-Hungary—when World War I began, and was promptly interned as a prisoner of war. But his studies would not be hindered: In his suitcase, he carried the papers on what he called "my great cosmic problem," and during his first night of confinement, he whipped out his fountain pen and turned to his calculations. "As I looked around my room after midnight," he recalled later, "I needed some time before I realized where I was. The little room seemed like the nightquarters on my trip through the universe."

Milankovitch did not stay long in his cell. Learning of his imprisonment, a Hungarian university professor who knew of the Serbian's accomplishments prevailed upon the authorities to parole Milankovitch to Budapest, where he could have access to the library at the Hungarian Academy of Sciences. There he spent the rest of the war years, developing a theory for predicting the earth's climate and completing a description of the climates of Mars and Venus. In 1920, his results were published in a work titled *Mathematical Theory of Heat Phenomena Produced by Solar Radiation,* in which the author demonstrated mathematically that widespread glaciation could be induced by astronomical changes that alter the amount and distribution of solar radiation reaching the earth. He also maintained that it was possible to determine the amount of radiation that had reached the earth at any time during the past. In short, Milankovitch was claiming he could prove that astronomical processes caused ice ages.

Among the many scientists who were impressed by Milankovitch's work

Puffing on a favorite pipe, Milutin Milankovitch works at his home in Belgrade, Yugoslavia, in 1950. After completing his astronomical studies in 1941, he spent the rest of his life writing his memoirs and a summary of his scientific work.

was the eminent German climatologist Wladimir Köppen, whose son-in-law, Alfred Wegener, had startled the scientific world in 1912 with his theory of continental drift. Now, Köppen and Wegener were in the process of writing a book about past climates. Invited to contribute to this project, Milankovitch readily agreed, and set out to plot a curve that would show the variations in radiation that he believed were responsible for the succession of ice ages.

James Croll had believed that variations in solar radiation at very high latitudes during the winter were the dominant factor in the onset of glaciation. But Milankovitch saw the matter otherwise. After lengthy correspondence with Köppen, he had become convinced that the decisive factor in glaciation is the diminution of summer heat in the temperate latitudes, not a reduction of winter radiation at the Poles—where temperatures even today are low enough to preserve a permanent snow cover. Working from morning until night, he drew curves showing how summer radiation in the middle latitudes—between lat. 55° N. and lat. 65° N.—had varied during the past 600,000 years. Finally, after 100 days, he finished his calculations and mailed the results to Köppen.

When the German scientist examined the work of his Serbian colleague, he was immediately struck by the marked similarity between the lines on the Milankovitch chart and the sequence of European glaciations established years before by the geographers Albrecht Penck and Eduard Brückner. Köppen informed Milankovitch that his astronomical theory had thus been confirmed, and asked him to attend a scientific conference to be held in Innsbruck, Austria. There, as Milankovitch listened from an inconspicuous last-row seat, Alfred Wegener presented a spirited lecture on continental drift and ancient climates, illustrating the section on the Pleistocene epoch with Milankovitch's painstakingly computed radiation curves. So well received was this new explanation for ice ages that Milankovitch slept that night "on a bed of laurels and soft pillows."

Köppen and Wegener included Milankovitch's work in their 1924 book, *Climates of the Geological Past,* and many geologists were convinced that the ice ages had at last been explained. Milankovitch, meanwhile, continued to elaborate and refine his theory, computing curves for latitudes both higher and lower than those that he had previously plotted. In 1930, he published his clearest statement yet on the causes of ice ages: *Mathematical Climatology and the Astronomical Theory of Climatic Changes.* In it, he demonstrated that radiation curves calculated for the higher latitudes are dominated by the 41,000-year tilt cycle, while curves for latitudes closer to the Equator are more heavily influenced by the 22,000-year precession of the equinoxes.

Aside from the fact that they corresponded with the assumed periods of glacial advances and retreats, Milankovitch's curves did not offer definitive proof that they delineated the causes of ice ages. But this correspondence seemed far too striking for mere coincidence. Scientists the world over came to accept the Milankovitch explanation for climate changes, and Milutin Milankovitch was convinced that his life's great work was done. For the first time since 1911, when he had set his lofty goal of scientific discovery, he was without a great challenge to face. "I am too old to start a new theory," he remarked wistfully to his son in 1941, "and theories of the magnitude of the one I have completed do not grow on trees." Ω

PROBING THE GLACIERS' PAST

The snows of the Ice Age lie entombed far below the surface of the Greenland and Antarctic ice sheets. Core samples that are drilled from these icy depths provide a detailed record of long-ago climate changes—and intriguing hints about their causes.

The difficulties of penetrating thousands of feet of polar ice are considerable. The ponderous, creeping motion of the ice tends to close coring shafts even as they are drilled, and the fluids used to keep the holes open sometimes freeze in the intense cold. But by 1982, scientists had managed to pierce all the way through the sheets of both Greenland and Antarctica.

Core samples extracted from the drill holes are scientific treasures, preserving continuous layered records of annual snowfalls going back perhaps 125,000 years. In ice that was formed recently, the annual accumulation layers manifest themselves as visible bands, allowing investigators to date the cores by counting the strata. For deeper cores, scientists rely on radiocarbon dating and on seasonal variations in oxygen-isotope ratios and dust content that define annual layers of snowfall.

Much of the cores' information is contained in the gases and dust trapped among the ice crystals. These whiffs of ancient air reveal that the Ice Age atmosphere had a low content of carbon dioxide—a gas that enables the atmosphere to retain the warmth of the sun—and a heavy burden of sunlight-screening dust, some of it volcanic. Both factors may have contributed to the Ice Age chill, but scientists are hesitant to rate their relative importance. Puzzling but tantalizing, the findings suggest that the deep ice contains many secrets yet to be unveiled.

A researcher lugs sample baskets toward a drill site on Antarctica's Ross Ice Shelf in a 1977 project that penetrated 1,375 feet to the sea water below the ice. Drill holes through land-based sections of the Antarctic ice sheet have pierced as deep as 7,100 feet.

Hot gases jet from the nozzle of a thermal drill being used to penetrate the Ross Ice Shelf. The drill was capable of cutting through the quarter-mile-thick shelf in nine hours, but the continuous creep of the ice made it necessary to rebore the hole every few days.

At a drilling site on the Greenland Ice Sheet, researchers remove a cylinder of ice from an aluminum core barrel. Another core sample, already sectioned and labeled, rests on the table.

Held against a strong light, a core sample
taken from Greenland ice displays dark layers
that were created by summertime melting.
The patterns of the annual bands are helpful to
scientists in dating the ice samples because
they hint at the severity of climate conditions
in the past: Narrower melt layers indicate
colder weather in the summer.

Working in a trench cut into the surface of the Antarctic ice sheet, a scientist slices a core of ice for future chemical analysis. The sterile garb that he is wearing prevents trace minerals on his skin and clothes from contaminating the sample.

Bathed in polarized light, the crystals of
a wafer of Antarctic ice glow in rainbow hues
(above). Polarized light enables scientists
to study the varying crystalline structures of
different samples, such as the compacted snow
just a few years old shown at right.

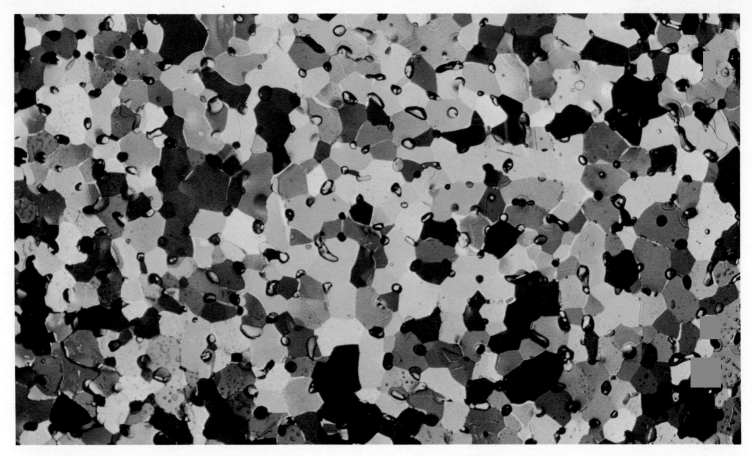

Air bubbles dot a sample of 500-year-old Antarctic ice, slightly magnified and photographed in polarized light. In the older, deeper samples, increased pressure has diffused the trapped air into the ice crystals, and the bubbles do not appear.

Ice deposited 19,000 years ago, near the height of the Ice Age, displays a band of tiny crystals formed around dust particles as they settled to earth after a volcanic eruption.

In a sample of 74,000-year-old Antarctic ice retrieved from a depth of 7,000 feet, the crystals have been combined and reshaped by the increasing temperatures and pressure near the bottom of the ice sheet. Individual crystals of golf-ball size have been found at such depths.

TESTING THE ASTRONOMICAL THEORY

I do not consider it my duty to give an elementary education to the ignorant, and I have also never tried to force others to accept my theory, with which no one could find fault." Thus, in the early 1950s, did a serenely confident Milutin Milankovitch dismiss those who did not accept his astronomical theory of the ice ages—and who were, in fact, beginning to find fault with it.

Some of the doubters were meteorologists, who complained that Milankovitch had considered only the amount of solar energy that reached different parts of the earth, ignoring the capacities of the atmosphere and oceans to store and circulate heat. Others were geologists, who could point to hard evidence that seemed to counter Milankovitch's claims. In 1953, for example, the German geologist Ingo Schaefer announced the results of his extensive studies of Alpine river terraces. Nearly half a century earlier, the German geographers Albrecht Penck and Eduard Brückner had examined the terraces, and their conclusions about the chronology of past ice ages had become a sturdy underpinning of the Milankovitch theory. Now, Schaefer had found fossils of warm-water mollusks in gravel layers that Penck and Brückner had insisted were laid down during an ice age. Many scientists downplayed Schaefer's discoveries, but it was becoming clear that the Milankovitch hypothesis was standing on suddenly shaky ground.

The astronomical theory's foundations were further weakened by the advent of radiocarbon dating, which enabled geologists to determine with increased precision the age of Pleistocene fossils. This revolutionary method had been developed by Willard F. Libby at the University of Chicago. Libby found that cosmic rays striking the atmosphere produce radioactive carbon atoms, which are eventually absorbed by animals and plants. The organisms take on this radiocarbon only while they are alive; after they die, their acquired radiocarbon begins to change into inert nitrogen. Libby discovered that this change takes place at a rate that is directly related to the age of the remains. After 5,568 years, half the radiocarbon in an organism's remains will have been converted into nitrogen; during the next 5,568 years, 50 per cent of the remainder will change. Since the rate of change is a constant 50 per cent over 5,568-year periods, Libby found that he could calculate the time of an organism's death by measuring the proportion of radiocarbon remaining in the fossil.

Previously, many geologists who embraced the astronomical theory had been content to date glacial debris—and thus establish a chronology of past

Among the oldest organisms on earth, bristlecone pines such as this gnarled veteran in California's White Mountains contain in their annual rings clues to climate conditions during the past 7,000 years. From such sketchy evidence researchers try to pinpoint the causes of global shifts in climate.

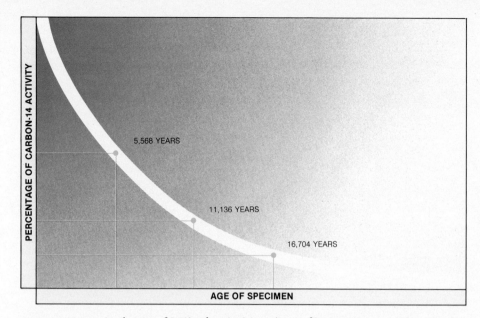

5,568 YEARS

11,136 YEARS

16,704 YEARS

AGE OF SPECIMEN

A logarithmic curve traces the consistent decay rate of radioactive carbon 14 over the millennia. The decay begins the moment any tissue dies. In 5,568 years, half of its carbon 14 disintegrates. Half of the remainder is lost in the next 5,568 years, and the decay continues at that rate until, after about 45,000 years, the amount of radioactive carbon becomes too small to measure—and hence date.

ice ages—on the basis of Milankovitch's solar-radiation charts. That is, the four Northern Hemisphere ice ages that had been deduced from glacial debris and other evidence were assumed to have occurred during the four most recent periods of low summer radiation delineated by Milankovitch. Armed with Libby's new radiocarbon technique, however, scientists were able to assign accurate ages to the materials left behind by Ice Age glaciers and to determine whether the glacial advances had indeed taken place during the times predicted by Milankovitch's theory.

Richard Foster Flint, a professor of geology at Yale University and an expert on the Pleistocene epoch, was among the first to apply radiocarbon dating to glacial events. Collecting wood, bones and other organic material that had been covered over by the Laurentide Ice Sheet as it plowed across eastern and central North America, Flint collaborated with geophysicist Meyer Rubin to demonstrate in 1955 that in most places the ice sheet achieved its greatest advance about 18,000 years ago, began to withdraw shortly thereafter and then hastened its retreat about 10,000 years ago.

According to the Milankovitch timetable, the most recent minimum in solar radiation occurred 25,000 years ago. Many scientists wondered why, if this were the case, the ice sheet had not reached its farthest extent until 7,000 years later. Milankovitch himself had anticipated such discrepancies, explaining that it would take a massive ice sheet some 5,000 years to react to alterations in the earth's radiation budget; an additional 2,000 years seemed to be a reasonable margin of error. But a new problem emerged when scientists studying the geology of Illinois found a layer of peat that they dated as 25,000 years old. It had been formed by the slow decay of vegetation that could have grown only during a time of relative warmth. Researchers soon found similar deposits of the same age in other parts of North America and in Europe as well, strongly suggesting that the Northern Hemisphere could not have been as cool then as Milankovitch said it was.

Again and again, geologists put the astronomical theory to the harsh test of radiocarbon dating and found the theory wanting. In the 1960s, postglacial logs and Roman-era bricks were identified in European gravel beds that Penck and Brückner had claimed were at least 20,000 years old. Most embarrassing of all, in an undisturbed layer of gravel that was said to date from the Ice Age, a Czech researcher uncovered a rusted bicycle part. The

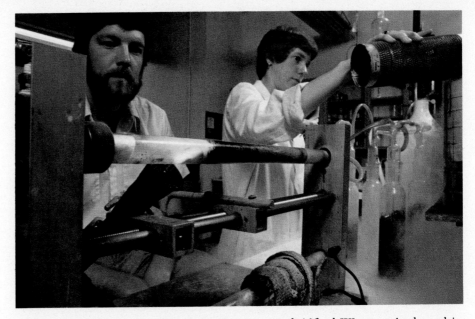

ice age chronology that Wladimir Köppen and Alfred Wegener had used in the 1920s to substantiate Milankovitch's hypothesis seemed shattered beyond all redemption.

Milutin Milankovitch, still convinced that he had uncovered the secret of the ice ages, died in 1958. Soon afterward, though a few supporters continued to argue his case, most of the world's scientists discarded the astronomical theory. And many of them offered other theories to fill the gap.

One widely discussed proposal was advanced in 1964 by the New Zealander Alex T. Wilson, a glaciologist who theorized that ice ages were caused by the ice sheets themselves. Wilson's scenario was relatively simple: As the Antarctic ice sheet grows thicker and heavier during thousands of years of snow accumulation, the weight borne by the underlying ice inexorably increases. The more pressure exerted on ice, the lower the temperature at which it melts, and eventually the bottom of the sheet reaches its pressure melting point; that is, it begins to melt even though it may be colder than 32° F. Since water under a glacier acts as a lubricant, Wilson proposed that at this stage the movement of the entire Antarctic ice sheet would accelerate, and the sheet would surge outward over the now-slippery polar bedrock and into the surrounding oceans. There, as it spread out and broke up, it would reflect increasing amounts of solar radiation back into space, causing a rapid cooling in the Southern Hemisphere; wind and marine currents would carry the chill northward, radically altering the world's weather patterns and causing ice sheets to form in the Northern Hemisphere. Over thousands of years, according to Wilson, some eight million square miles of North America and Eurasia would be buried under ice.

Wilson went on to tackle the problem of how such an ice age might end. Again he found his answer in the behavior of the ice itself. The massive outward surge of the colder ice from the central dome cools the base of the ice sheet below its pressure melting point again; deprived of the lubricating effect of water, the ice would stop surging into the sea, the floating ice shelves would be diminished by melting and the solar energy they had been reflecting would begin to warm the southern oceans. The heat would be transferred to the northern oceans and would in turn warm the continents of the Northern Hemisphere. As the temperature rose, the American and Eurasian ice sheets would melt, and the earth would emerge from an ice age.

Very neat. Not only did the theory sound plausible, it offered an explanation for the cyclic nature of ice ages. Studies of surging glaciers show that after the massive transfer of ice from a high elevation to a lower one where melting is more rapid, a glacier will retreat until enough new snow accumulates to trigger a fresh surge. Bruarjokull, an outlet glacier of Vatnajokull, the icecap in southeastern Iceland, surged in the late 19th Century and then retreated for 70 years before surging again in 1963; records show that its surges have occurred every 70 to 100 years for the past 350 years. A more striking periodicity has been demonstrated by the Vernagt Glacier in the Ötztal Alps of Austria. Since 1599, the Vernagt has surged four times, at intervals of 79, 93 and 74 years. (A surge in about 1927, which could have been expected from the average interval of 82 years, failed to materialize, and the glacier has not surged since.)

It seemed reasonable to suppose that an ice sheet as huge as Antarctica's would expand and contract at a much slower pace—perhaps at intervals that coincide with the sequence of Pleistocene glacial and interglacial periods. But the evidence in support of Wilson's theory is not wholly convincing. No one has actually witnessed an Antarctic surge, and the likelihood of such an event occurring in the future can only be inferred from mathematical calculations—which indicate that the bottom ice at the center of the sheet is close to its pressure melting point—and field studies showing that some of the sheet's inner portions have already reached the pressure melting point.

In 1970, British scientists Hubert Lamb and Alastair Woodroffe advanced what became known as the "snowblitz" explanation of ice ages. Instead of the catastrophic surging of the Antarctic ice sheet envisioned by Wilson, the two Britons suggested that an ice age could be started by nothing more dramatic than a succession of harsh winters. Suppose, they said, that the high latitudes of Europe suffered several years of severe winters followed by cool summers. If the snow cover on the hills and mountains were deep enough to last through the summers, the increased reflectivity of the land surface would reduce air temperatures, and more precipitation would fall as snow and less of the snow would melt. The initial worsening of weather would be amplified and would continue until ice sheets developed on the lowlands in the high latitudes.

History suggests that the snowblitz theory is less than foolproof. In medieval times there did indeed begin a series of abnormally cool winters and summers that lasted for several hundred years; but although the deteriorating climate led to an expansion of Northern Hemisphere glaciers, warmer conditions soon returned.

Other theories raise the possibility that variations in the earth's atmosphere might start and stop ice ages. Most of these theories fall into two categories, one involving a reduction in the efficiency of the atmosphere's radiation traps—which retain radiant energy in the atmosphere—and the other involving an increase in the atmosphere's reflectivity.

Of the heat energy radiated from the surface of the planet, only about one third escapes into space; the rest is absorbed by gases in the atmosphere and then reemitted. About 42 per cent of the retained heat is trapped by water vapor and carbon dioxide. These substances create what is known as the greenhouse effect: Like the glass in a greenhouse, they allow incoming short-wave solar radiation to pass through freely but absorb the long-wave energy radiated from the earth's surface. Though water vapor accounts for most of the trapped radiation, it does not seem likely that the absence of water vapor will ever cause an ice age. The British astronomer Fred Hoyle has pointed out that the earth's average temperature of 59° F. is more than warm enough to support the rate of evaporation necessary to maintain the water-vapor trap. If temperatures did decrease significantly, then conceivably the amount of atmospheric water vapor might fall to a critically low level—but in that case, the destruction of the water-vapor trap would be a result of a drastic change in climate, not its cause.

Carbon dioxide is a far more efficient heat trap than water vapor; while carbon dioxide constitutes only .03 per cent of the volume of the atmosphere, it absorbs 15 per cent of the energy radiated by the earth. Complete removal of the carbon dioxide trap would reduce the average temperature of the earth to 27° F. This calculation suggests that even a slight loss of atmospheric carbon dioxide might plunge the planet into an ice age and has prompted a number of rather dramatic scenarios. In one, a series of volcanic eruptions and forest fires increases the amount of carbon dioxide in the atmosphere, initially leading to a brief warming of the climate and to increased plant growth. But since the additional vegetation absorbs more carbon dioxide, the volume of the gas present in the atmosphere subsequently falls, causing a cooling trend that is accentuated by a corresponding decrease in the efficiency of the water-vapor radiation trap. The earth enters an ice age. But with the lowering of global temperatures and the advance of great ice sheets, much of the world's vegetation dies, releasing carbon dioxide into the atmosphere. The radiation traps are restored, temperatures rise and the ice age ends.

Ingenious as the model sounds, it oversimplifies. Plants do, indeed, fix carbon dioxide, but the amount they absorb from the atmosphere is slowly compensated for by the release of the gas from the oceans and other sources—among them the digestive processes of the world's termite population, which is thought to produce more than twice as much carbon dioxide as does the combustion of fossil fuels. It would take the loss of almost all the carbon dioxide in the atmosphere to cause an ice age, a loss that would also cause a catastrophic reduction in plant growth and a consequent obliteration of animal life. There is no evidence of such a sequence of events having occurred at the onset of an ice age.

Of the several theories involving increased atmospheric reflectivity, one has commanded attention ever since Benjamin Franklin proposed the idea in the 18th Century. This is the notion that periods of cooler weather can be initiated by volcanic eruptions. Although a dramatic increase in volcanic activity might at first have a short-term warming effect on the atmosphere by increasing the carbon dioxide level, its net result would be to cool the earth by throwing up a vast quantity of dust that would reflect sunlight and also act as the nuclei for the condensation of clouds.

The climatic consequences of volcanic eruptions have been gauged. In 1815, for example, Mount Tambora, a volcano on an island east of Java, blew up with spectacular violence, ejecting some 25 cubic miles of debris into the atmosphere—enough to plunge an area of thousands of square miles into darkness for three days.

Nearly 150 years later, two American scientists, Henry and Elizabeth Strommel, delved into the meteorological records of America and Europe to assess the volcano's impact. They found that during the year following the

Mount Tambora eruption, New Haven, Connecticut, had experienced the coldest June on record, while the summer in the Swiss city of Geneva had been the coldest since 1753. The New England harvest of 1816 had been disastrous, with yields of the staple crops of corn and hay drastically reduced by the poor growing season. Among the locals the year was remembered as "Eighteen Hundred and Froze to Death."

Further evidence of the climatic effects of volcanic eruptions was collected after the 1883 eruption of Krakatoa. The titanic explosion of this Indonesian island volcano, audible 2,200 miles away in Australia, left 1,000 feet of water where there once had been land and destroyed 300 towns. On the other side of the world, in the French city of Montpellier, meteorologists found that solar radiation reaching the earth's surface was diminished by 10 per cent for three years after the eruption. Another devastating volcanic outburst, on Bali in 1963, was followed by a drop in solar radiation of 5 per cent at Pretoria, South Africa. Yet none of these cataclysmic events produced a long-term lowering of global temperatures, let alone an ice age. In each case, the volcanic dust forming the reflective shield was dispersed in no more than 13 years; the reservoir of heat contained in the oceans is capable of maintaining the earth's temperature for about this length of time, so the long-term effects on global climate proved negligible.

Considerably more research will be required before the exact relationship between volcanism and ice ages can be determined. According to expert opinion, intense volcanic activity would have to be sustained for at least 1,000 years to produce an ice age. Researchers have, in fact, found evidence that numerous volcanic eruptions occurred at the onset of the last ice age, but few scientists would suggest that they could have been its sole cause.

Not surprisingly, the search for the primary cause of ice ages has led to close scrutiny of the sun itself, the source of all but an infinitesimal fraction of the earth's heat. Although the sun seems a paragon of reliability, its output varies in response to violent processes that take place within its seething mass.

Solar variation has been linked to the waxing and waning of sunspots—areas of disturbed activity, visible to the naked eye at sunset, that appear darker than the rest of the sun's surface because they are cooler. Since the 17th Century, when telescopes were invented, scientists have observed cycles of sunspot activity lasting about 11 years. They have also postulated longer ones of 89 years and 178 years, and guessed that other, even longer cycles probably occur. Climatologists have tried to link sunspot cycles with climatic change, but apart from a possible correlation between sunspot activity and increased rainfall in some parts of the globe, the evidence is inconclusive.

The risk of attaching too much importance to the effects of sunspot cycles was wryly summed up in 1973 by British scientist J. W. King, a man whose dedication to his work was equaled only by his passion for the sport of cricket. According to King, information contained in the *Wisden Cricketers' Almanack*—the enthusiasts' bible—"can be used to show that, of the 28 occasions on which cricketers have scored 3,000 runs in a season in England 16 have been in sunspot maximum and minimum years; the five years in which this rare phenomenon occurred more than once were all sunspot minimum or maximum years. Likewise, 13 of the 15 occasions on which a batsman has scored 13 or more centuries in a season

A Volcano's Chilling Legacy

Among the possible causes of the ice ages, none has been demonstrated more dramatically in historic times than the climatic changes wrought by great volcanic eruptions.

Benjamin Franklin was perhaps the first to surmise that volcanoes might affect climate. In 1784 he proposed a connection between a cataclysmic eruption in Iceland the summer before and a haze that had enveloped the earth and so weakened the sun's rays that the following winter, Franklin noted, "was more severe than any that had happened for many years."

With the eruptions of the Mexican volcano El Chichón between March 28 and April 4, 1982, another climate-altering cloud began to spread around the globe. By most standards, the explosions at El Chichón were modest. But because of atmospheric conditions, the chemical composition of the magma, and the fact that the blasts were directed straight upward, the eruptions launched a high-altitude cloud of ash and gas at least 20 times more massive than the cloud released by the 1980 eruption of Mount St. Helens.

The cloud consisted mostly of sulfur dioxide gas. Ash particles quickly coalesce and settle to the ground after an eruption, but the gas remains suspended for years, gradually diffusing around the globe in a layer 10 to 18 miles above the earth. The sulfur dioxide reacts with atmospheric moisture to form a sunlight-veiling miasma of sulfuric-acid droplets—a "universal fog," as Franklin termed it in 1784.

Scientists calculated that the Mexican eruptions would lower average temperatures during the following winter by about half a degree. While such an effect is hardly the harbinger of a new ice age, some experts believe that hundreds of eruptions over many centuries could trigger another relentless advance of the world's ice sheets.

As a villager hurries past, a column of gas and ash rockets skyward during an eruption of Mexico's El Chichón on April 2, 1982. A larger eruption two days later lofted gas and debris to an altitude of more than 18 miles.

Images recorded by a weather satellite on April 4, 1982, show a small, circular burst of gas and ash directly over El Chichón (*left*) blossoming into a cloud hundreds of miles across (*center*) in only six hours. Five hours later (*right*) winds had swept the debris almost to Cuba.

Scientists in Wyoming (*below*) fill an unmanned research balloon with helium for a flight into the upper atmosphere. With particle counters slung beneath them, balloons such as the one shown at right penetrated the layer of debris from El Chichón to measure the density of the globe-girdling cloud.

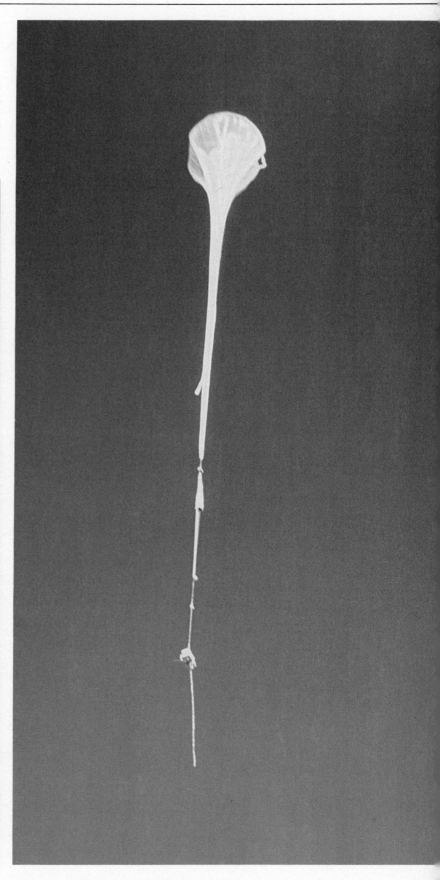

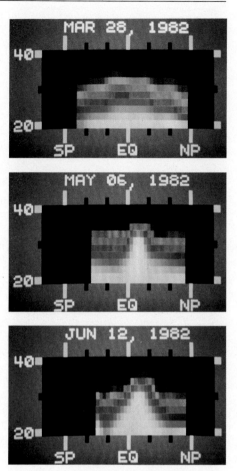

Computer-generated profiles of a Pole-to-Pole slice of the stratosphere, based on infrared readings taken by satellite just before the eruptions of El Chichón *(top)* and at monthly intervals afterward *(middle and bottom).* chart the spread of the volcanic emissions. By June 12, the cloud was still more than 18 miles high and had spread into the Southern Hemisphere.

Photographed from the slopes of Hawaii's Haleakala Crater, the crystalline sunlight of August 1980 *(top left)* contrasts with the diffuse and milky sun of July 1982 *(left).* after El Chichón's eruptions veiled parts of the earth with a high-altitude haze.

took place in, or within a year of, a sunspot maximum or minimum year."

King's point is a telling one; when scientists seek to correlate such real or theoretical occurrences as sunspots, snowblitzes, volcanoes, surging ice sheets and atmospheric changes with the chronology of past ice ages, the results are hardly more satisfactory than when cricket scores are used. It has become increasingly apparent that the climate changes that lead to an ice age are brought about by several kinds of factors operating in different ways and over a wide range of time scales. Even if all the factors could be isolated and their impacts assessed, a battery of powerful computers would have to work for many decades to calculate the overall effect on the climate of a particular time and place.

And yet, as it turns out, one theory fits the ideal, precisely matching its posited cause of glaciation with the timetable of the Pleistocene ice ages. Ironically enough, it is the same astronomical theory that was proposed by Joseph Adhémar in 1842 and by James Croll in 1864, and then refined in the early decades of the 20th Century by Milutin Milankovitch.

Even while most scientists had turned their backs on the astronomical theory to seek other explanations for the onset of ice ages, some had continued to search for evidence that would corroborate Milankovitch's painstakingly charted timetables of the variations in solar radiation. Frustrated by the difficulties of working on land, where the glacial record has to be compiled piecemeal from many different sites, geologists looked for places where conditions have been stable right through the Pleistocene ice epoch and where there are enough sensitive—and datable—indicators to provide an uninterrupted history of climatic events. They found them at the bottom of the oceans, where the floors of the abyssal plains are covered with thick beds of sediment laid down at more or less constant rates.

Scientists have been plumbing the ocean depths ever since 1872, when H.M.S. *Challenger* set out from England on a worldwide research voyage. The ship returned after more than three years at sea, bearing such a wealth of data that the results of the cruise were not fully compiled and published until 1895, when investigators released the last volume of a 50-volume report. One of the myriad observations concerned foraminifera—tiny planktonic organisms informally called forams that build chambered shells as they mature. Forams, the report noted, are found in all the world's oceans, some species only in warm waters, others only in cold waters.

For those who would later seek to unravel the mystery of the ice ages, this was a highly significant finding; it meant that an examination of fossil foram sequences on the seabed could indicate whether the ocean was warm or cold at the time that the creatures died and sank to the bottom. Indeed, James Croll himself had observed years before that a record of climate—and thus of the succession of ice ages—could be compiled by examining the skeletons, shells and other remains that lie in "the deep recesses of the ocean, buried under hundreds of feet of sand, mud, and gravel."

Unfortunately, it would be many years before scientists came up with the technology required to bring up sea-floor core samples that were long enough to permit a reading of climate over extended periods of time. Meanwhile, they did the best they could, raising cores—most of them only a yard or so long—with heavy sections of pipe dropped into layers of sediment. Studying cores taken in this way from the Atlantic floor in the mid-

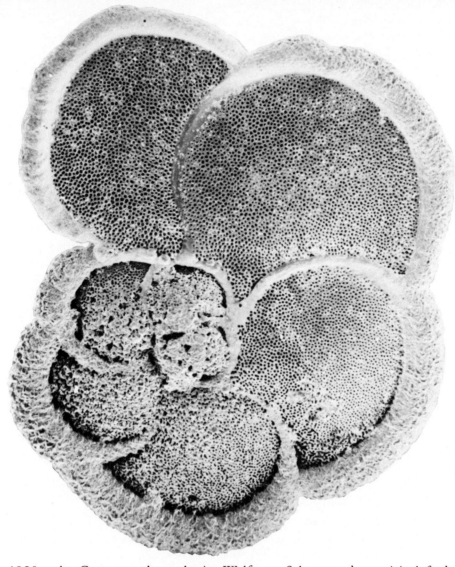

Fossils of marine microorganisms have provided scientists with an important indication of past climatic changes. For example, *Globorotalia menardii*—shown here magnified 80 times—lives only in warm waters: Thus, its presence in sea-core sediment layers indicates prehistoric warm periods; its absence suggests a cooler climate, like that of an ice age.

1920s, the German paleontologist Wolfgang Schott made a critical finding. He recognized in the sediments three distinct layers with different populations of forams. The uppermost layer, laid down in the recent geological past, contained a high concentration of warm-climate species, including *Globorotalia menardii*. The second, older layer was richer in cold-climate types, and *menardii* was completely absent. But in the oldest of the three layers, *menardii* was back again, together with a high proportion of other warm-climate species. Schott deduced that the layer devoid of *menardii* was deposited during the last ice age, when the Atlantic was cooler, while the other two strata were laid down during the preceding and present interglacials.

Before any sweeping conclusions could be drawn, however, scientists needed to study much longer cores, a prospect that was made possible in 1947 when the Swedish oceanographer Björe Kullenberg devised a tube that could collect 45-foot cores—containing materials deposited over hundreds of thousands of years—by sucking up sediments at the same time that it was being driven into the seabed. In the early 1950s, research ships began bringing up such samples, many of which were collected for Columbia University's Lamont Geological Observatory. When David B. Ericson of the Lamont staff examined them, he found, as had Schott before him, that layers of warm-water *menardii* remains alternated with those of the cold-water forams. Ericson was convinced that the layers with *menardii* present indicated interglacial periods and the others marked an ice age.

Geochemists at Lamont determined that the boundary between the topmost, *menardii*-rich layer and the second layer had been formed rapidly, some 11,000 years ago. Ericson noted that the timing of this rapid change from cold to warm climate coincided with the dates that had been deduced from radiocarbon dating of glacial debris on land. In a paper describing their findings, Ericson and his colleagues noted that "further correlation of events both in the ocean and on land during this interval may lead to an understanding of some of the factors causing glaciation."

Meanwhile, other scientists were conducting a parallel line of research that involved chemical analysis of fossil forams. The method they used had been suggested in 1947 by Harold Urey, a Nobel laureate at the University of Chicago. It consisted of measuring the ratio of two oxygen isotopes—atoms that are virtually identical but different in atomic weight—that are absorbed from sea water by the shells and skeletons of marine organisms. Urey and his associates had found that organisms from cold water contained a higher proportion of the heavier isotope, designated oxygen 18, or O-18, than did organisms living in warmer water. The remains of the warm-water creatures had a higher ratio of the lighter isotope, known as O-16.

In the 1950s, the Italian-American geologist Cesare Emiliani applied Urey's theories to eight deep-sea cores. After radiocarbon-dating the upper sections of the cores and then estimating the rates of sedimentation, Emiliani decided that there had been no fewer than seven complete glacial-interglacial stages during the past 300,000 years and that they had occurred in a time sequence that agreed fairly well with the variations predicted by Milankovitch. In its broad outlines, Emiliani's work also agreed with Ericson's findings, although there were some major differences; certain periods that Ericson's foram analysis had identified as warm were shown by Emiliani's methods to have been cold.

So spirited was the debate over the contradictory findings that in 1965 the National Science Foundation held a special conference to try to settle the dispute. John Imbrie, then a professor of geology at Columbia University, attended the meeting, and later told the story of the controversy and its aftermath in his book, *Ice Ages: Solving the Mystery*. At the conference, Imbrie pointed out that Ericson and Emiliani had all but ignored the possibility that factors other than temperature may cause variations in foram concentrations. Imbrie decided then and there to develop an analysis technique that took into account such things as water salinity and the amount of food available, as well as water temperatures in winter and summer.

At a meeting held in Paris in 1969, Imbrie announced the results that he and his associate, Nilva Kipp, had obtained when they studied a Caribbean core with this multiple-factor technique: Whereas Emiliani's research indicated that surface water temperatures in the Caribbean had dropped by almost 11° F. during the last ice age, Imbrie's multiple-factor method showed a drop of only 3.5° F.

Furthermore, when the core inspected by Imbrie and Kipp—a core that had been examined much earlier by Ericson—was analyzed for oxygen-isotope ratios, the zones that Ericson had identified as cold were shown to be warm by both the isotope and multiple-factor methods. "Apparently," Imbrie wrote, "some environmental factor other than surface water temperature (but often correlated with it) caused *Globorotalia menardii* to appear and disappear cyclically in deep waters of the Atlantic Ocean."

It was a pleasant Friday afternoon when Imbrie spoke. As he relates in his book, the attractions of Paris were proving more alluring than the pursuit of science, and he had an audience of two—one of them unable to understand English, the other a young British geophysicist named Nicholas Shackleton. Talking together after the lecture, Imbrie and Shackleton realized that their independent work on the problem had led them to the same answer: Changing ratios of oxygen isotopes in marine fossils are caused primarily by fluctuations in the size of ice sheets, not by variations in sea temperatures. Their tentative conclusion was based on the fact that because O-18 is heavier than O-16, water molecules containing O-18 do not evaporate as readily; therefore, water rising from the oceans in the form of vapor and subsequently falling as precipitation contains a smaller proportion of O-18 than do the oceans themselves. If water deficient in O-18 were to be locked up on land in the ice sheets, the proportion of the heavy isotope in sea water would rise, and this increase would be reflected in the ratios of the oxygen isotopes present in forams and other marine organisms.

In a few years, this theory would be put to a test—with ultimate results that Imbrie and Shackleton could hardly have dreamed of on that balmy afternoon in Paris.

Meanwhile, other scientists in various fields had been working on ice age timetables, aided by an advanced dating technique—accurate for as far back as 150,000 years—that measures the decay rate of uranium contained in calcium carbonate, a substance found in such materials as limestone, coral and mollusk shells. Many of these chronologies seemed to validate the Milankovitch radiation curve. In 1965, for example, geochemist Wallace S. Broecker reported some interesting findings that he and some colleagues had made when they dated fossil coral reefs in the Florida Keys and the Bahamas. Since coral can grow only at certain depths, it provides an accurate record of former sea levels. Broecker's studies indicated that the sea had stood much higher 120,000 and 80,000 years ago, presumably during periods of warm climate when vast amounts of water had been released from the melting ice sheets. Noting that present sea levels are also considerably higher than they have been at times of great glaciation, Broecker observed that these three known periods of high sea levels closely correspond to the warm periods calculated by Milankovitch in his radiation curve for lat. 65° N.

Soon, other researchers were reporting similar findings elsewhere. Brown University geologist Robley K. Matthews, for example, investigated the terraced coastlines of Barbados and determined that the steplike terraces had been formed by the growth of coral reefs at former sea levels. According to his calculations, the age of one terrace reef was 80,000 years, that of another was 125,000 years—a near-perfect match with the findings of Wallace Broecker. But Matthews found something quite different, too: a middle terrace that indicated a time of high sea level about 105,000 years ago.

Sadly for believers in the astronomical theory, the Milankovitch curve did not show a radiation maximum in that time period—not, at least, at lat. 65° N., where the effects of axial-tilt variations space the radiation peaks at intervals of some 41,000 years. But when Broecker—curious about the seeming anomaly of the 105,000-year-old Barbados shoreline—examined additional Milankovitch curves, he found that those for lower latitudes showed peaks corresponding to all of the dates assigned to the three Barbados terraces. At these latitudes, it seemed, the precession cycle's

Clues in Coral Seas

Coral reefs, built by living creatures, preserve a detailed record of changes in the weather and the environment. The growth of coral, like that of trees, is accomplished with a succession of annual layers. Each year, coral polyps—tiny marine animals—deposit a layer of light-colored limestone, which commonly ranges from ¼ inch to ½ inch in thickness, followed by a thin layer of denser, darker growth most easily detected in X-ray photographs.

For optimum growth, coral depends on sunlight, good water quality and favorable temperature. When unfavorable environmental conditions, such as severe cold or, possibly, water pollution, stunt the growth of the coral, an abnormal dark band appears in that year's limestone accumulation. Such "stress bands" not only document the ravages of weather, they also enable researchers to cross-date samples from different coral outcroppings.

By 1982, scientists had used samples of living coral to follow this trail of climatic evidence several hundred years into the past. One 10-foot-long core, drilled from a reef off the Florida Keys, preserved a 360-year record of storms and cold fronts—along with traces of global air pollution from industry and radioactive fallout from nuclear weapons testing.

Fossil corals may extend the growth-ring record dramatically; coral growth rings remain discernible in limestone as old as 360 million years. Although samples that old cannot be dated precisely, a sequence of coral records spanning tens of thousands of years may one day be assembled. When X-rayed and analyzed for the oxygen isotopes that indicate the amount of ice present in the global environment, the fossil corals may chronicle the advance and retreat of ice sheets with an accuracy never before achieved.

Steadying a hydraulic drill as it chews into a coral mass on a Philippine reef, a diver extracts a core sample to be used in a study of the effects of offshore oil wells on coral growth.

A scientist removes a sample of coral from a diamond-toothed coring bit. To prevent marine organisms from attacking the living coral through the sample hole, the researchers will plug the opening with a cement cap.

The annual growth of a coral is revealed in this X-ray photograph of a wafer-thin section, cut lengthwise from a core sample. Each thin, dark band marks the end of a year's growth.

Limestone terraces on the north coast of Papua New Guinea record the dramatic sea-level changes of the last 125,000 years. The ages of the terraces, established by scientists in 1974, helped to confirm the dates of interglacial warm periods cited by Milutin Milankovitch in his astronomical theory of recurring ice ages.

22,000-year period—the time it takes for the earth's axis to wobble in a complete circle in space—was influential enough to modulate the effects of axial tilt. Reef terraces in Hawaii and New Guinea yielded similar data, indicating that past periods of high sea level could indeed be explained by application of the astronomical theory of ice ages.

On a different scientific front, researchers were seeking to refine geological chronologies by matching switches in the magnetic polarity of undersea sediments. Tiny iron particles in most rocks become permanently magnetized in alignment with the earth's magnetic poles at the time the rocks are formed, and scientists had recently found that the same process occurs in ocean sediments. The phenomenon of magnetic reversal—which is probably caused by disturbances in the earth's molten core—had first been noted in 1906 by Bernard Brunhes, a French geophysicist who discovered that the iron-rich particles in an ancient lava flow had been magnetized so that the north and south magnetic poles were interchanged. During the 1960s, scientists using the recently developed potassium-argon dating technique—which measures the rate at which a radioactive potassium isotope found in rocks changes to an isotope of argon—determined that the earth had reversed its magnetic polarity a number of times during the past four million years. The last switch took place about 700,000 years ago. All

deposits laid down since this occurrence, which marked the start of the period called the Brunhes Epoch, have "normal" magnetic polarity; strata laid down during the 300,000 years prior to the event show reverse polarity; and before that the orientation was normal.

Since deposits from land and sea and from all parts of the world contain the same magnetic records, the identification of the times of the reversals is a means of correlating the geological chronologies estimated by different methods in different regions. Older deposits revealed that a magnetic switch took place about two million years ago. This is a significant date, for it corresponds roughly to the time that geologists assign to the beginning of the Pleistocene epoch and the start of the glacial-interglacial cycle that has continued to the present.

Now, armed with a way to date sea-floor sediment cores that contained evidence of past climate changes, scientists would be able to determine whether previous cold and warm periods coincided with the cycles predicted by Milutin Milankovitch. But such proof would also have to explain the fact, deduced from continuing studies of sediments on land and on the sea floors, that the 22,000- and 41,000-year cycles—which Milankovitch had believed to be most critical to radical changes in climate—seemed to be superimposed over longer cycles of 100,000 years, a figure reminiscent of James Croll's theory that variations in the earth's orbital eccentricity are paramount in bringing about climate changes. It appeared that the great Pleistocene ice ages had developed slowly in cycles of about 100,000 years; after a number of oscillations, each ice age had come to an abrupt end. Only if Milankovitch's shorter cycles could be related in some way to these 100,000-year periods could his astronomical theory be accepted as an explanation of the cause of the ice ages.

In the spring of 1971, as part of the International Decade of Ocean Exploration, a group of scientists and researchers organized a series of studies known as CLIMAP—the Climate Long Range Investigation, Mapping and Prediction project. One of their first missions was to analyze sea cores and deduce the climate changes that have taken place during the 700,000-year Brunhes Epoch.

To achieve the goal, investigators needed a core rich in forams that could be analyzed for oxygen isotopes. In December, CLIMAP scientists located such a specimen—it had been raised from the western Pacific early in the year—and after confirming that it dated back beyond the magnetic reversal that marked the start of the Brunhes Epoch, they shipped samples of the core to Nicholas Shackleton at Cambridge University.

Shackleton, an expert at analyzing the isotopic contents of marine fossils, studied the core samples and plotted two isotopic curves, one showing the ratio of light and heavy oxygen isotopes in the remains of surface-dwelling forams, and the other plotting isotopic variations in forams that lived on the sea floor. If, as Cesare Emiliani had theorized some years earlier, the proportion of oxygen isotopes in marine fossils is governed by sea temperatures, the second curve should have shown much smaller deviations than the first: No matter what the climate, the temperature of the water at the bottom of the ocean remains close to freezing. In fact, as Shackleton showed the CLIMAP team in mid-1972, the two isotopic curves were nearly identical.

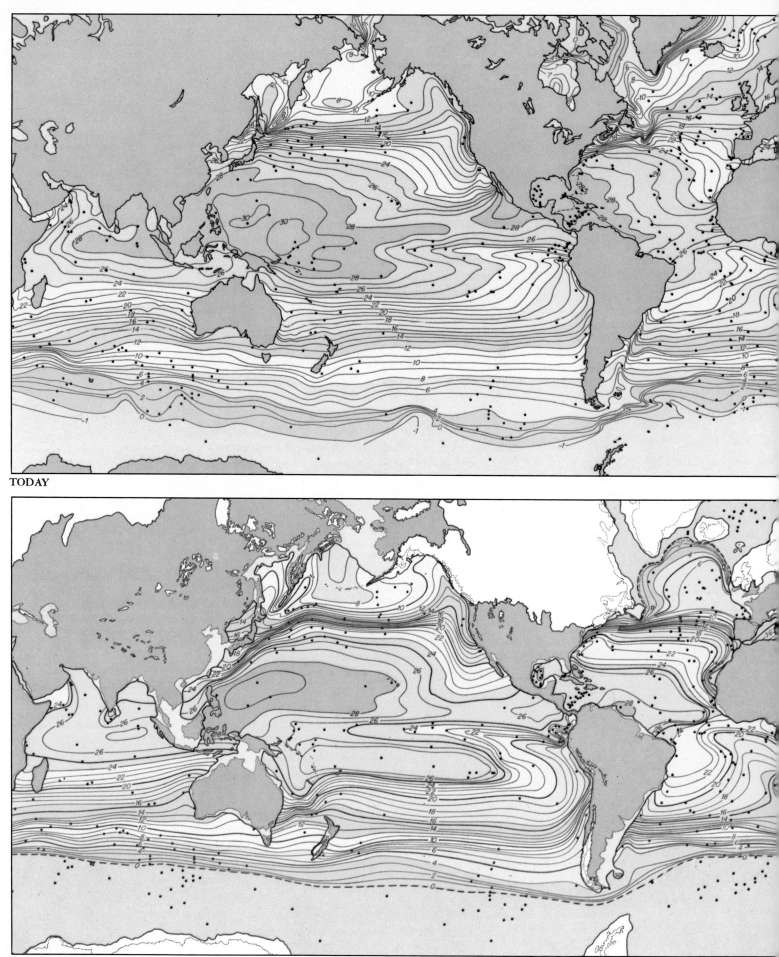

TODAY

18,000 YEARS AGO

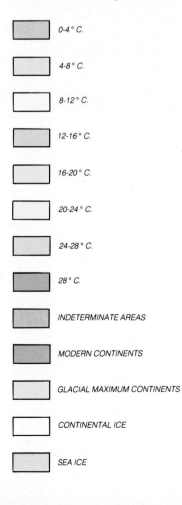

Color-coded maps, prepared by the CLIMAP climate study group, chart the surface temperatures of present-day oceans *(top)* and the estimated temperatures 18,000 years ago *(bottom)*, at the height of the Ice Age. The Ice Age temperatures were determined by studying core samples of ocean-floor sediment at locations indicated by dots on the maps.

0-4° C.

4-8° C.

8-12° C.

12-16° C.

16-20° C.

20-24° C.

24-28° C.

28° C.

INDETERMINATE AREAS

MODERN CONTINENTS

GLACIAL MAXIMUM CONTINENTS

CONTINENTAL ICE

SEA ICE

It was just as Shackleton and John Imbrie had surmised in Paris three years before. Imbrie would write later that both of Shackleton's curves "reflected changes in the proportion of light isotopes in the ocean—not changes in water temperature. And, because sea water was mixed rapidly by currents, any chemical change in one part of the ocean would be reflected everywhere within a thousand years. All along, Emiliani's curve had been a chemical message from the ancient ice sheets. When the glaciers expanded, light atoms of oxygen were extracted from the sea and stored in the ice sheets—altering the isotopic ratio of oxygen in sea water. When the glaciers melted, the stored isotopes flooded back into the ocean, returning it to its original composition." In short, the two curves were not directly indicating changes in climate; instead, they documented a consequence of climate change—the waxing and waning of glaciers, the comings and goings of the ice ages.

Shackleton's painstaking analysis yielded additional revelations. The core showed a definite sequence of 19 stages of warming and cooling over the past 700,000 years, making it possible for scientists to estimate the duration of each stage. More significantly, there were clear indications that major climate changes had occurred at intervals of some 100,000 years, the same time period suggested by the notion that climate was affected primarily by changes in the eccentricity of the earth's orbit around the sun.

This 100,000-year cycle was so dominant on Shackleton's curve that he could not determine whether the less-pronounced fluctuations reflected the 41,000-year axial-tilt cycle and the 22,000-year-old precession cycle. Milankovitch's astronomical theory of the ice ages remained unproved. Before long, however, another CLIMAP researcher, James D. Hays of Columbia University, clarified the situation by looking at two sediment cores from the southern Indian Ocean—one that had been raised in 1967 and one that had been brought up in 1971. While the cores' records did not extend all the way back to the start of the Brunhes Epoch, they went far enough—450,000 years—to provide a sufficient time span for valid analysis. Moreover, the sediments had built up more rapidly than those in Shackleton's Pacific core, providing a thicker accumulation for each cycle—and thus a more detailed account of the climate changes that had occurred.

When Hays and Shackleton examined the evidence from the Indian Ocean cores, they found clear imprints of the 100,000-year cycle. They also saw unmistakable signs of the shorter cycles of 41,000 and 22,000 years. "We are certain now," they announced, "that changes in the earth's orbital geometry caused the ice ages. The evidence is so strong that other explanations must now be discarded or modified."

Earlier conclusions drawn from radiocarbon dating—which at first seemed to invalidate the Milankovitch theory—had already been modified considerably. As geological knowledge expanded, researchers realized that a slight waning of the ice sheets occurred some 25,000 years ago, indicating that climates would have been warm enough to produce such apparent anomalies as a 25,000-year-old peat layer in Illinois. So the evidence obtained by radiocarbon dating no longer stood in the way of the astronomical theory.

Still, not everyone was convinced that the astronomical theory was the answer; some scientists maintained that the degree of solar-energy variation is too small to account for ice ages. The British astronomer Fred Hoyle, for

A yawning crater 600 feet deep and almost 4,000 feet in diameter scars the Arizona desert where a meteorite struck thousands of years ago. Some scientists have suggested that much larger meteorites colliding with the earth could throw enough dust and debris into the stratosphere to trigger an ice age.

example, pointed out that while the 4 per cent warming produced by the so-called Milankovitch effect has been used to explain the melting of the Northern Hemisphere ice sheets some 13,000 years ago, the current 3 per cent warming in the Southern Hemisphere is making little impression on the Antarctic ice sheet. He also claimed that the cycles were too long and ponderous in their effects to cause the sudden shifts in climate that apparently marked the beginning and end of the Pleistocene ice ages.

In 1981, Hoyle advanced an intriguing theory of his own, a hypothesis involving giant meteorite strikes and the peculiar properties of super-cooled water droplets. When air containing tiny water droplets is cooled progressively, Hoyle explained, the water can remain liquid down to $-40°$ F., at which point it turns into what polar explorers call diamond dust—tiny crystals of highly reflective ice. Suppose, Hoyle argued, that the water vapor in the upper atmosphere were chilled from its present $-4°$ F. to the critically low $-40°$ F. In such a case, he claimed, a veil of diamond dust precipitated from a vapor layer no thicker than .01 millimeter would reflect almost all incoming sunlight back into space, plunging the earth into an ice age within a few decades.

According to Hoyle, a strike by a giant meteorite would produce the necessary cooling in the upper atmosphere. If the meteorite were made of

138

reflective stone, he suggested, the dust particles generated by its destruction would cast a pall around the earth, screening out the sun's rays long enough for the atmosphere to cool to the point where diamond dust formed. Heavier dust particles would eventually fall to earth, but the tiny ice crystals would remain—keeping the earth in ice-age conditions—until the temperature of the upper atmosphere rose above −40° F. In Hoyle's scenario, the end of an ice age is heralded by the impact of another meteorite—this one, composed primarily of metal, would throw up a cloud of radiation-absorbing particles. The diamond dust would melt in a flash, while the metallic particles suspended in the atmosphere would absorb heat and cause the earth's temperature to rise enough to melt the ice sheets as quickly as they had formed.

The Hoyle meteorite hypothesis seems ingenious enough, and scientists have, in fact, estimated that the planet has been hit by about 5,000 giant meteorites of more than half a mile in diameter over the past 600 million years. But the theory has a number of flaws, not least among them the extreme improbability that huge meteorites would have smashed to earth at the regular intervals that have marked the beginning and end of global glaciations. Nor is it likely that the effects of the debris thrown up by the impact of two different types of meteorites would be so totally dissimilar. Indeed, of all the theories that have been proposed, the astronomical theory remains the only explanation that can be clearly linked to known chronologies of the Pleistocene ice ages. Even so, the nature of climate is so complex that while astronomical events have doubtless been the main triggers of the ice ages, it is likely that other phenomena—among them volcanic eruptions, sunspots and perhaps even meteorite strikes—have also conjoined in some way to help turn the earth from warm to cold and back again.

Deep-sea sediments contain climatic records that extend back some 50 million years, and the geological evidence of the last Pleistocene ice sheets is clearly visible in many parts of the world. Thus much more is known about the glaciations of these comparatively recent times than about those that took place in the more distant geological past. But scientists have long known of other ice ages scattered far back in geologic time. Much of the evidence for these ancient ice ages comes from tropical and subtropical regions that were largely untouched by the most recent glacial advances. As long ago as the 1850s, for example, British geologists came upon traces of pre-Pleistocene glaciation in tropical India. And by the turn of the century, geological signs of ancient ice sheets had been uncovered in Australia, South Africa and Brazil.

At first, the only explanation seemed to be that glaciers had in fact overwhelmed the whole world, from the Poles to the Equator. Then, in 1912, the German meteorologist Alfred Wegener—who later did so much to publicize and support Milankovitch's astronomical theory—announced his theory of continental drift. The continents, Wegener proposed, are not fixed and stable, but drift like rafts on the planet's fluid mantle. For more than 50 years, most geologists scorned Wegener's heretical notions. By the late 1960s, however, enough supporting evidence had been found to vindicate his hypothesis and to revolutionize the earth sciences. Today, few scientists doubt that the continents, borne on enormous tectonic plates of the earth's crust, are slowly but constantly moving across the face of the planet.

A sampling of cosmic-dust particles—remnants of comets, asteroids and other space debris—found on the sea floor 600 miles east of Hawaii glint in a jewel-like display at the California Institute of Technology. Such particles provide a record of astronomical events that may have contributed to the advent of ice ages.

By studying the magnetic orientation of ancient rocks, scientists can determine a continent's position, relative to the magnetic poles, at the time the rocks were formed. And by noting the differing orientations of rocks of different ages, they can chart the continents' past movements over the globe. Such paleomagnetic evidence did much to confirm the theory of continental drift, and it also accounted for the paradoxical evidence of glaciation near the Equator: The areas in question had been located much closer to the polar regions when ice sheets covered them. Indeed, it has since become clear that glaciation of sufficient scale to be called an ice age can occur only when large parts of the earth's land surface are located near the Poles or in high latitudes.

The theory of continental drift led to one of the most remarkable discoveries in ice age studies. During the 1960s, scientists analyzed the magnetic orientation of rocks from many parts of the world and concluded that North Africa had been located over the South Pole during the Ordovician period, about 450 million years ago. If they were correct, there should be traces of ancient glaciation in the Sahara. At about the same time, French petroleum geologists working in southern Algeria stumbled on a series of giant grooves that appeared to have been cut into the underlying sandstone by glaciers. The geologists alerted the scientific world and assembled an international team to examine the evidence. The team saw unmistakable signs of an ice age: scars created by the friction of pebbles incorporated into the base of glaciers; erratic rocks that had been transported from sources hundreds of miles distant; and formations of sand typical of glacial outwash streams.

One of the scientists, Rhodes Fairbridge of Columbia University, described the effect on the team as "electrifying," and went on to observe: "Here we were privileged, beneath the hot Sahara sun, to see the detailed record of a giant glaciation, precisely dated, and just where it had been predicted to be by the evidence of the paleomagnetists. Our French hosts were not unprepared for the occasion. There was a refrigerator on our supply vehicle, and out of it miraculously emerged a bottle of the finest champagne, ice cold. And so we drank to the health of the discoverers, to the visitors, and to the Ordovician ice age!" Ω

SIGNS OF A LONG SAHARAN WINTER

Covering an area of some 3.5 million square miles in northern Africa, the Sahara is the world's largest and most forbidding desert. Across much of its expanse, rainfall averages less than five inches per year, hot, dry winds lash the barren landscape almost constantly, and summer temperatures can exceed 130° F. in the shade.

The Sahara has not always been so torrid, however. As recently as 5,000 years ago, the region was dotted with large, shallow lakes and covered with vegeta-tion. And long before that, the Sahara was overrun by ice sheets that left behind unmistakable imprints—gouged bedrock, clutters of erratic boulders, cliffs of glacial debris.

The glaciation occurred some 440 million to 465 million years ago, during what geologists call the Ordovician period. At the time, Africa, Australia, Antarctica, India and South America were joined in the supercontinent known as Gondwana, which surrounded the South Pole. The location of this great land mass in a polar region permitted ice sheets to form; they pushed steadily into the midlatitudes, waxing and waning during millions of years.

Then, about 200 million years ago, Gondwana started to break apart, and the fragments began moving relentlessly across the face of the world toward their current positions as the modern continents. The climate in formerly frigid lands changed considerably, but time could not eradicate the geological clues left behind by the ice.

When ice sheets engulfed the Sahara in the Ordovician period, Africa—then joined to other continents—was centered on the South Pole *(inset)*. Present-day sandstone cliffs, perched atop granite basement rock on the Plain of Admer *(below)*, were formed by sand deposited when the ice finally melted.

SOUTH POLE

Glacial striations—gouged by sharp rock
fragments embedded in the base of an ice sheet—
mark the bedrock of southern Algeria. In
some parts of the Sahara, such grooves can be
traced for hundreds of miles.

A member of a geological expedition inspects the remains of a water scour, a deep channel that was cut by the rushing waters of a melting ice sheet. Later, as the flood subsided, the channel was filled in with waterborne sand.

145

Erratic boulders—some bearing glacial scratch marks—litter a Saharan hillside. The hill itself is composed of till—materials such as clay, sand and gravel that were also left behind by a retreating ice sheet.

Starkly outlined against the desert sands, this esker — formed by deposits laid down by water flowing through a channel at the edge of a melting ice sheet — snakes for about 30 miles across the western Sahara.

THE NEXT ICE AGE

The interglacial period that has seen the rise of human civilization and the disappearance of the megafauna is ending, and there can be little doubt that the great glaciers of the Ice Age will return. One stark statistic suggests our jeopardy: The four previous interglacials lasted between 8,000 and 12,000 years, and the present one, called the Holocene, has already endured a little longer than 10,000 years.

These figures amount to a clear warning, but their predictive value should not be overestimated. André Berger, a leading theoretician of celestial mechanics who has calculated in painstaking detail the variations in the earth's orbit during the past million years and for 60,000 years to come, has concluded that, if the progression continues as it did during the Pleistocene, the earth will be well into an ice age between 3,000 and 7,000 years from now. On the other hand, the early stages of climatic cooling could begin much sooner; the British climatologist Hubert Lamb has said that pronounced cooling within the next two centuries is not out of the question.

The impact of such an event on the modern world cannot be conjured. At the current pace of technological and cultural change, little can be said with certainty about the capabilities of the human species two centuries hence, let alone 30. Yet the consequences of imposing on the temperate latitudes, with their current concentrations of the world's population, industry and agriculture, the environment of a present-day Lapland or Arctic Canada, would unquestionably be profound.

Unfortunately for humankind's peace of mind, the number of processes involved in the triggering of an ice age, and the relationships among them, approach the infinite. Few of these processes are understood fully enough to permit a confident projection of their behavior, let alone of their cumulative effects.

For instance, in 1976 the American physicist Johannes Weertman calculated that the earth might experience a perpetual ice age, without even the grace periods of interglacials, if the Northern Hemisphere land masses—North America, Eurasia and Greenland—were positioned 300 miles farther north than they are now. Even the warmest summer resulting from the changing relationships between earth and sun would not provide enough heat to melt the annual accumulation of snow and ice. Instead, it would build up year after year and century after century in a permanent glaciation. But if these land masses were instead to drift 300 miles to the south and the oceans were to occupy their former positions, Weertman reckoned, every

Silhouetted against a wall of tightly packed Antarctic snow, glaciologists mark off the distinct annual layers of accumulation. Gauging the variations in polar snowfall from year to year helps scientists to understand past changes in global climate and predict future trends.

summer would be warm enough to prevent the growth of any ice sheets.

These scenarios suggest the difficulty of making any forecast of an ice age. Although Milutin Milankovitch's astronomical theory now seems secure, there is much uncertainty about the mechanisms that translate orbital change into climate changes. Some of the answers surely lie in what scientists call "proxy data" about the remote past—microfossils, volcanic dust, pollen and other chemical and physical traces of ancient climates left in ocean sediments, polar ice and peat bogs.

Records of more recent climatic shifts also promise to be useful: An understanding of the relatively minor variations between the colder and warmer intervals of the present interglacial may assist in the reconstruction of earlier, more drastic changes. In addition, precise data on fluctuations in solar output, which may be linked to climate, are now being gathered by satellite-borne instruments. And glaciologists are studying the ice sheets of both Greenland and Antarctica for clues to long-term trends. Analysis of the mountains of data being teased from these varied sources should eventually permit at least a highly educated guess about the advent of the next ice age.

Flowing past the southeastern United States toward the British Isles and Scandinavia, the waters of the Gulf Stream appear in violet in this computer-enhanced satellite image. During the Ice Age, cold waters from the Arctic moved south to deflect the course of the warm current toward northern Africa.

The sea-floor cores that provided the supporting data for Milankovitch's astronomical theory remain a primary source of information. Analysis of corebound evidence—fossils of surface-dwelling plankton and of deep-sea foraminifera, oxygen isotopes, and particles of sand and clay—has allowed two geophysicists, William F. Ruddiman and Andrew McIntyre of Columbia University's recently renamed Lamont-Doherty Geological Observatory, to outline the role the North Atlantic might have played in the growth and decay of the North American ice sheets.

The scientists focused on two periods—the start of the last ice age, some 115,000 years ago, and a particularly frigid episode that occurred about 40,000 years later. At both of these times, summer sunshine was at a minimum in the Northern Hemisphere because the tilt of the earth's axis had reached its smallest angle. With less solar radiation reaching the middle and high latitudes, low temperatures prevailed during the summers, and less of the winter snow that fell on the plains, plateaus and mountains from Labrador northward was melted. Even as the land was beginning to chill, however, a different condition prevailed in the oceans. The kinds of fossil plankton in sea-floor sediments show that the ocean off the Canadian coast was as warm as it is today, averaging around 65° F. The Gulf Stream, warmed by the stronger sunshine in the southern latitudes, flowed northward past the already ice-rimmed coast of maritime Canada. The oxygen-isotope records in sediment cores suggest that the cooling of the oceans lagged behind that of the land by some 3,000 to 5,000 years.

Paradoxically, the warmth of the North Atlantic waters actually hastened the growth of the ice sheets, according to Ruddiman and McIntyre. For one thing, warm water evaporates faster than cold water, and so the maritime air near the frigid continent was laden with moisture—ready to fall, under the right conditions, as snow. And the very contrast in temperature between land and ocean could have supplied the right conditions by creating a storm track that channeled winds originating over the ocean northward across Canada. When the moist oceanic air was chilled over the land, its water vapor condensed and fell as snow. The ocean was too warm to freeze over in winter, and so it offered its ready supply of moisture year round.

The reflectivity of the growing continental mass of snow and ice amplified the cooling effects of declining solar radiation. The Canadian glaciologist Roy M. Koerner has examined the reflectivity, or albedo, of the present-day icecap of Canada's Devon Island, which lies inside the Arctic Circle. He found that dry snow on the island's highest, coolest elevations reflects as much as 85 per cent of the sunlight striking it. This reflected energy escapes into space without contributing any significant amount of heat to the earth's atmosphere. During ice-sheet growth, the expanding area of cool continental air reacts with the incoming flow of oceanic moisture to produce heavy snowfalls over larger and larger areas, as the snow line creeps downward from high plateaus to lower elevations.

A particular kind of sediment Ruddiman and McIntyre found in the cores documents the warmth of the North Atlantic at the onset of glaciation. These deposits consist of sand and clay particles larger than those carried to sea by ocean currents or wind. The debris was scraped from the surface of the earth by the ice sheets, carried to sea by icebergs that broke free from the glaciers' edges, and then released when the icebergs succumbed to warm water. The location of the deposits, called ice-rafted detritus, shows that when the ice sheets were in the early stages of growth, the icebergs that calved from them melted near the coasts of Greenland and Newfoundland. In later periods, icebergs drifted as much as 1,000 miles farther south, to the latitude of Spain, before meeting water warm enough to melt them.

The ratio of the heavy oxygen isotope O-18 to the lighter O-16 in the North Atlantic cores suggested to Ruddiman and McIntyre that the ocean became so cold about 6,000 years after the continental ice sheets first began to form that vast expanses froze over in winter. With a lid of sea ice in place for part of the year, the principal moisture supply for the ice sheets was largely sealed off, and their rate of growth dropped rapidly. Even in summer, when much of the sea ice melted, the Atlantic remained so cold that evaporation was limited. Thus, at the depth of the Ice Age, some 18,000 years ago, the size of the continental ice sheets stabilized for a time.

Oxygen-isotope profiles and the fossil content of ocean cores indicate that when the last two ice ages terminated—the earlier one some 127,000 years ago and the most recent about 10,000 years ago—the oceans were still remarkably cold. Just as they had lagged behind the continents in growing cool, they took longer to warm as the ice sheets began to melt. During both periods of ice-sheet decay, the axis of the earth was in an orbital configuration that brought maximum solar radiation, or insolation, to the Northern Hemisphere in summer—and minimum insolation in winter. The North Atlantic continued to freeze over during the bitterly cold winters and provided comparatively little moisture to nourish the ice sheets. It took several thousand years for the waters of the North Atlantic to warm to their present temperatures. They are now, Ruddiman and McIntyre note, evaporating fast enough to feed the growth of the ice sheets should the amount of summer insolation fall below a critical threshold.

Sediments deposited while the ice sheets were in retreat disclose that the ocean surface was almost barren of microscopic plankton. Ruddiman and McIntyre think this may have been due to a deluge of fresh water—the runoff from ice sheets melting rapidly in the summer. Floating on the heavier salt water, the layer of fresh water changed the salinity of the Atlantic's surface so much that plankton were obliterated; they did not return until the meltwater had dispersed, several thousand years later.

Although sediments deposited on the sea floor contain some hints about the composition, temperature, and turbulence of the earth's atmosphere during ice ages and interglacials, cores drilled from the great ice sheets that survive in Greenland and Antarctica provide much more detailed information. These sheets—composites of annual layers of snow compressed and transformed into ice—provide an unbroken record stretching hundreds of thousands of years into the past. The age of the oldest of this ice has not yet been determined, but scientists think that just above the bedrock of East Antarctica lies ice that was formed 500,000 years ago.

Perhaps most important to understanding the mechanics of ice ages are the ice sheets' component water molecules, which contain the oxygen isotopes that indicate past changes in temperature. But the ice sheets also contain airborne particles—dust, volcanic debris, sea salts, and various isotopes formed in the atmosphere—that fell on the snow when it was fresh. In addition, the layers of ice are riddled with tiny air cavities—remnants of the atmosphere just as it was when the snow fell. These ancient air samples and the other constituents of the ice sheets (with the exception of

Ancient Ice in an Andean Lake

While studying South American flamingos in 1977, California biologist Stuart Hurlbert was startled to find in a remote Andean lake ice islands as much as a mile long and rising about 20 feet above the water's surface. Carbon-14 dating of sediments within the ice indicated that this may be the oldest lake ice in the world outside of polar regions.

The islands owe their survival to a covering layer of aragonite, an intensely white form of calcium carbonate that reflects most of the sunlight striking it. Somehow frozen into the ice, aragonite sediments now provide a foot-thick layer of insulation that has protected these remarkable ice islands for as long as 7,000 years.

A gap in the insulating cover of aragonite exposes the strata of an ancient ice island in the Andes. The dark layers are composed of pure fresh-water ice; the alternate layers contain light-colored minerals—chiefly aragonite.

radioisotopes, which decay with time) remain perfectly preserved as long as they are frozen.

Much as the isotope content of a layer of sea-floor sediment does, the ratio of the heavy O-18 isotope to the lighter O-16 in a layer of ice indicates the temperature at the time it was formed, but with a very important difference. A large proportion of O-18 in sea sediment indicates a colder climate, but just the opposite is true in ice. Since more heat energy is required to vaporize water molecules containing the heavier isotope, a higher proportion of O-18 in an ice layer means that the air temperature was relatively high when that water evaporated from the ocean and later fell as snow. Thus, snow that falls in summer has a higher O-18 content than winter snow. On a much longer time scale, the snows of an interglacial are richer in this isotope than snows deposited during glaciation.

Ice cores have one very significant advantage for scientists trying to establish and refine the chronology of climatic shifts. The top inch or so of sea-floor sediment is frequently stirred by bottom-dwelling creatures; consequently, the layers cannot be dated with precision. But ice layers are not likely to have ever been disturbed by living creatures. Because the concentration of the O-18 isotope generally peaks in summer, declines in winter, then peaks again, the ice between two peaks represents a single year's snowfall. Painstaking measurements of O-18 levels—in areas where the tem-

peratures are so low all year that the ice layers have not been muddled by melting—have identified the ice formed each year as far back as 1000 B.C.

Willi Dansgaard, a Danish geochemist who was one of the first scientists to study ice cores, thinks it will be possible to assign precise dates to ice layers up to 7,000 years old. Because O-18 has a tendency to diffuse through the ice, the annual peaks and valleys become blurred in layers older than that. They can be dated, though with less exactness, by the radioisotopes they contain—principally carbon, beryllium and chlorine.

Of all the cores removed from the Greenland Ice Sheet, the one containing the most ancient ice is the Camp Century core, named for a research outpost located in the northwestern part of the island. Camp Century is one of four major study sites for scientists working with the Greenland Ice Sheet Project, a cooperative venture sponsored by Danish, Swiss and American institutions. The deepest, and hence oldest, layer of the Camp Century core once lay 4,600 feet below the surface of the ice sheet and was probably laid down some 125,000 years ago, before the advent of the Ice Age. Dansgaard's analysis of the O-18 content of the bottom 1,000 feet of the core yielded details of the climatic history of Greenland—and, by inference, of the earth—from the end of the last interglacial to the end of the subsequent Ice Age, some 10,000 years ago. The temperature trends signaled by the O-18 levels parallel the changes indicated by sea-floor cores from the Indian Ocean and the North Atlantic.

The Camp Century core shows that the climate of Greenland became sharply colder, and the Ice Age began there, about 120,000 years ago. The

154

Left: In an array of plastic greenhouses on Canada's Ellesmere Island, University of Toronto researchers are developing ways for inhabitants of Arctic regions to grow fresh vegetables despite ice-age conditions. *Above:* A shirt-sleeved scientist inspects a seedling plot in one of the sun-warmed enclosures.

onset of the cold was soon followed by heavy annual snowfalls for the next 5,000 years. (Dansgaard thinks the heavy snowfalls account for the rapid build-up of the ice sheets and the consequent drop in sea level.) But these Ice Age snow samples contain almost as much O-18 as snow that had fallen thousands of years before, when the climate was milder. Thus the Camp Century core confirms that the North Atlantic near Greenland did indeed remain warm, as Ruddiman and McIntyre concluded from their research, long after the ice sheets had begun to grow.

The Camp Century core also records another drop in temperature and an ice advance some 75,000 years ago. This timing coincides with that calculated by Ruddiman and McIntyre from North Atlantic sediments—and with the period when, according to the astronomical theory, the Northern Hemisphere was having the cool summers needed for ice-sheet growth.

Particles of clay, volcanic dust and sea salts embedded in ice cores indicate that during the last glaciation, especially toward its end, the atmosphere was turbulent and dirty. In the Camp Century core, Ice Age layers contain 12 times as many such particles as do layers formed during the interglacial that followed. According to Ellen Mosley-Thompson and Lonnie Thompson, a team of glaciologists at Ohio State University's Institute of Polar Studies, the clay particles came from the great expanses of coastal plains bared by the receding oceans. Stiff winds whipping across these flats picked up the dirt and distributed it around the globe. Greenland ice layers tend to be twice as dirty as their Antarctic counterparts; the Ohio State researchers think the reason is simply that Greenland lies closer to more sources of particles, since the Northern Hemisphere has much more land than the Southern. The same strong winds that bore the clay particles also picked up droplets of water as they swept across the ocean, and thus they account for the high concentrations of sea salts found in Ice Age cores.

The volcanic dust that often accompanies the clay and salt deposits provides strong evidence that frequent and violent eruptions accompanied the Ice Age. Many scientists think that the large quantities of volcanic dust circulating in the atmosphere might have made the climate even colder, since the particles would have reflected solar radiation away from the earth. It may even be that the Ice Age was prolonged by the dust.

Besides being stormy and dirty, the Ice Age seems to have had scant snowfall. Around 15,000 years ago, the annual accumulation at Camp Century, as shown by chemical analyses, was only one third to one half the average winter's snowfall there now. This evidence supports the picture developed by Ruddiman and McIntyre after their study of sea-floor sediments—of a cold, ice-covered ocean from which little moisture evaporated.

Geologists Chester C. Langway and Michael M. Herron have turned up evidence that when the Ice Age finally ended, the climatic transition came about very rapidly—within a few decades or even less. Segments of two cores, one from Camp Century and the other from the Dye 3 research station in southern Greenland, reveal that the concentration of windborne sea salts fell by about 75 per cent in no more than a century. Apparently, the harsh winds that characterized the Ice Age for thousands of years had died away in the moderating climate of the interglacial.

Another fascinating change that accompanied the end of the Ice Age involved the amount of carbon dioxide in the atmosphere. Scientists are particularly interested in this gas for two reasons: For one thing, it is inti-

The Dwindling Snows of Kilimanjaro

Within 200 miles of the Equator, patches of ice have persisted since the Ice Age atop three East African mountains. The most famous of these remnants consists of the snows of Kilimanjaro in Tanzania —which are actually glaciers. Roughly 1.5 square miles of ice crown the 19,340-foot summit, where the temperature stays below freezing throughout much of the year. Mount Kenya and the Ruwenzori mountain range on the border of Zaire and Uganda also have small icecaps.

Odd though it seems, equatorial Africa experienced at least four major glaciations during the Pleistocene. Many mountains in the region are marked by moraines—piles of rocky rubble deposited by glaciers. The moraines on Kilimanjaro indicate that glaciers at one time extended almost halfway down the mountain's flanks. At its coldest, however, the climate of Pleistocene Africa was relatively mild, with average temperatures throughout the continent only about 6° F. lower than today's.

With the end of the Ice Age, the African glaciers began to shrink. They probably would have melted away hundreds of years ago were it not for the advent of the Little Ice Age—the period of cold between the 15th Century and the mid-19th Century—during which they maintained their size and may even have advanced slightly. Around 1800, the glaciers began to shrink once more. Between the late 19th Century and the mid-20th Century, about one quarter of the snows of Kilimanjaro melted. By the end of the 20th Century, if the climate does not turn colder once again, they will disappear entirely.

Right: Thirty-foot-high ice terraces dwarf climbers on Kilimanjaro's Northern Glacier. The fluted surfaces, shaped by meltwater streams and solar radiation, betray the glacier's steady deterioration.
Below: The mountain's highest peak lifts a beacon of gleaming ice over the Amboseli Plain, where year-round temperatures average 85° F. Glaciers on its lower peak had melted by the 19th Century.

mately linked to the metabolic reactions of plants, which release it as waste, and so can serve as an index of biological activity. Second, it is a key component in the so-called greenhouse effect—the mechanism by which the atmosphere captures the warming infrared waves emitted by the earth's surface. Because carbon dioxide absorbs infrared energy, the temperature of the atmosphere depends in part on how much of the gas it contains.

Analysis of air bubbles trapped in three ice cores—from Camp Century in Greenland and from Antarctica's Byrd Station and Dome C—has provided a chronology of atmospheric changes from 40,000 years ago up to the present. During the last major glaciation, the amount of carbon dioxide in the atmosphere fell drastically, by about 25 per cent, reaching its lowest level during the last 2,000 years or so of the Ice Age. Why this occurred is not known for certain, but the decline may reflect a reduction of plant life on the icy continents and in the surface waters of the ocean. Then, at about the time the interglacial began, carbon dioxide became more abundant. Scientists have not yet pinpointed whether the carbon dioxide level increased before or after the climate changed, but the rise certainly would have helped warm the earth. During the centuries-long transition from glacial to interglacial climate, average global temperatures rose about 4° F. A group of scientists from Switzerland's University of Bern have speculated that an intensified greenhouse effect might have accounted for a third of the temperature increase. But they attribute a larger percentage of the increase to the fact that, as the ice sheets melted, the newly exposed earth absorbed far more solar energy than the ice had.

The average global temperature continued to rise for thousands of years. John E. Kutzbach, a meteorologist at the University of Wisconsin, has calculated that during the month of July 9,000 years ago, the earth received about 7 per cent more solar radiation than it does during July in present times, and the temperature averaged about .4° F. higher. The Northern Hemisphere was experiencing very warm summers because two orbital factors were reinforcing each other: The earth was making its closest approach to the sun during the Northern summer, and the hemisphere was tilted sharply toward the sun, so the high latitudes were receiving maximum insolation. Besides being warm and sunny, the summers were wet, Kutzbach thinks, with about 8 per cent more precipitation than in modern times. Parts of northern Africa and northwest India that had been arid during the Ice Age became very humid and rainy; in some spots, the precipitation rate may have been double what it is today. The ancient shorelines of some tropical lakes show that early in the interglacial their waters stood 100 feet or more above present levels. And human evidence attests to a different environment. Saharan cave drawings from the period show elephants, giraffes, and even river-dwelling crocodiles and hippopotamuses.

Temperatures peaked around 4000 B.C. and remained stable for about 2,000 years. During this period—called the climatic optimum, for its benign conditions—many regions were about 5° F. warmer on the average than they are today, according to calculations based on pollen distribution and the oxygen-isotope ratios in the Greenland ice cores. The melting polar icecaps had released so much water by this time that sea level was 300 feet higher than it had been at the end of the Ice Age. Civilizations were flourishing in regions of the world that today are deserts; in northwest India, for instance, a people called the Harappans spread from the Indus

Valley, raising grain on well-watered plains and building numerous towns.

After 2000 B.C., temperatures in the Northern Hemisphere began a slow decline. Drought struck in tropical and subtropical areas. The rains failed in the Indus Valley, and the Harappans abandoned their fields and towns to the encroaching sand dunes. In Egypt, winds piled sand and soil in the dried-up beds of Saharan streams and lakes that for millennia had supplied farmers with water, and the Nile's annual flood level dropped sharply. An Egyptian named Neferty overstated the case somewhat when he bemoaned a drought that had shriveled the Nile. "The River of Egypt is empty," he wrote; "men cross over the water on foot."

While the lower latitudes were experiencing drought, Europe north of the Alps became increasingly cold and wet. Glaciers advanced in mountainous regions, and forests were transformed into bogs. In North America, the Paleo-Eskimos abandoned their high-latitude Arctic hunting grounds and migrated south to Labrador and the Hudson Bay, while glaciers formed in the Rockies south of what is now the Canadian border, for the first time since the Ice Age. Around 450 B.C., temperatures began to rise again, reaching a peak around 1000 A.D. Ever since that time, although they have oscillated up and down, the overall trend has been toward cooler temperatures. The most extreme downward turn occurred around 1500, when a cold period known as the Little Ice Age set in. This neoglacial event, as climatologists call it, persisted into the 19th Century.

The implication of the long-term cooling trend over the past millennium—that the global climate may already have begun its descent into the next glaciation—is supported by the astronomical theory. Changes in the earth's orbital configuration since 6,000 years ago, when the temperatures of the interglacial peaked, have been in a direction that favors the growth of ice sheets. The axis of the earth is not tilted so sharply as it was then, so summers are cooler; they should become cooler still, as the angle of tilt continues to decrease. In addition, perihelion—the earth's closest approach to the sun—now occurs in January, rather than in summer. According to the astronomical theory, this arrangement means still-lower temperatures (and even less melting of winter snows) during Northern Hemisphere summers, and relatively mild weather during the winters, when insolation is at its peak. Such a combination of factors favors the growth of ice sheets, since mild winter temperatures increase the air's capacity for moisture and hence the abundance of snowfall, and cooler summers permit more of the snow to persist long enough to be converted into glacier ice.

Whether the earth has begun to play out the scenario of a returning ice age remains to be seen, and before definite conclusions can be drawn, scientists must know much more about the current state of the ice masses of Greenland and Antarctica. To track changes in the great polar ice sheets, glaciologists need better measurements of their present dimensions, especially their thickness. Current estimates of the volume of the ice sheets may be off by as much as 50 per cent because of the lack of accurate data.

The best measurements of the thickness, extent and surface contours of ice sheets have been made by radar altimeters, carried aboard aircraft since the late 1950s and on satellites since 1972. By 1982, about 10 per cent of Antarctica and 50 per cent of Greenland had been mapped from satellites operated by the U.S. National Aeronautics and Space Administration (NASA). The altimeters they carried were sensitive enough to measure elevation changes of as little as two inches; by comparing those surveys with data from future satellite flights, glaciologists will be able to detect even minor thickening or thinning of the ice sheets. Similar maps have been made of the ice shelves—the flat, floating masses of ice that fringe parts of the Antarctic and Greenland coastlines. By 1982, enough measurements had been made for the United States Geological Survey to begin drawing finely detailed maps of the ice shelves to serve, like the information on ice-sheet thickness, as base lines for detecting future changes.

Scientists also want to keep track of the West Antarctic marine ice sheet, which rests on bedrock lying as much as 3,300 feet below sea level. At present, ice shelves shield the margins of the marine ice sheet from direct erosion by the sea. But if the ice shelves should disintegrate, this ice sheet would be prone to sudden collapse and disintegration—a process that might take less than 200 years. The icebergs produced by such a breakup would drift toward the midlatitudes, perhaps cooling the ocean waters and increasing albedo enough to trigger a new glacial advance.

The collapse of a marine ice sheet is one of the stochastic, or random, events that scientists think could sharply accelerate the transition from an interglacial to a glacial climate. Another example would be an unusual

Two geologists, mapping the southern part of Antarctica's Victoria Land by snowmobile, head for a rocky outcrop. Behind the sled, a bicycle wheel linked to an odometer measures the distance from one field camp to another.

160

outburst of volcanic activity for a number of years, which would inject large quantities of ash into the atmosphere, where it would block incoming solar radiation. The category also includes a far less spectacular phenomenon—a succession of several years of unusually cold weather.

Oscillations that Willi Dansgaard and his colleagues have found in the O-18 content of the Camp Century ice core suggest that such cold spells have occurred approximately every 180 years for the past thousand years. Why this happens is a subject of speculation. Variation in the solar constant—the amount of solar radiation reaching the earth—is one possibility. Precise measurements of the solar constant are difficult to obtain from earth because of the interference of the atmosphere, but the launching of NASA's Solar Maximum Mission satellite in 1980 solved this problem. Orbiting some 300 miles above the earth's surface—where the atmosphere is extremely thin—Solar Max has detected only minor fluctuations in the solar constant, amounting to a few tenths of 1 per cent over periods of a week or so. Changes of that magnitude are too small to cause significant swings in temperature on earth; scientists estimate that it would take a reduction of one half of 1 per cent or so in the sun's brightness over several years for the average global temperature to fall 1° F. It is too early to draw

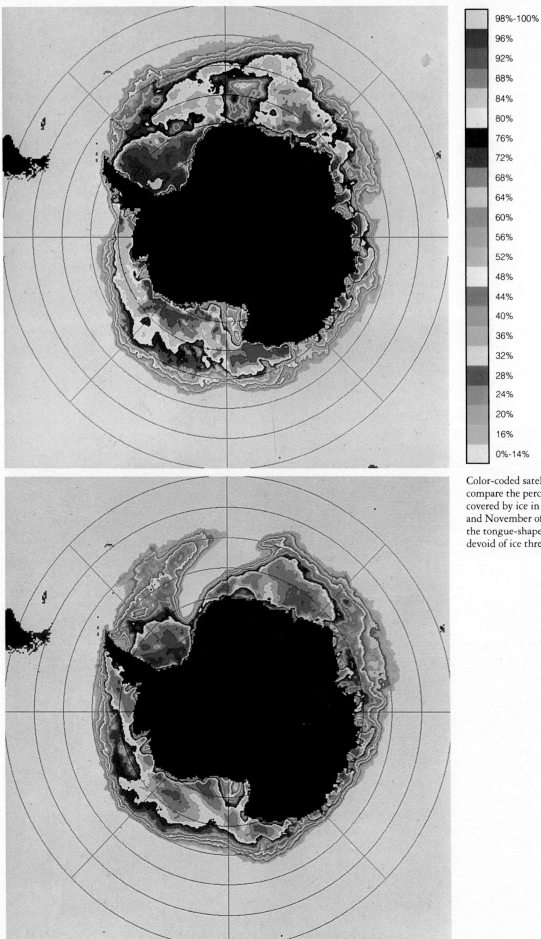

	98%-100%
	96%
	92%
	88%
	84%
	80%
	76%
	72%
	68%
	64%
	60%
	56%
	52%
	48%
	44%
	40%
	36%
	32%
	28%
	24%
	20%
	16%
	0%-14%

Color-coded satellite images of sea ice compare the percentages of Antarctic waters covered by ice in November of 1973 *(top)* and November of 1976. Frozen over in 1973, the tongue-shaped Weddell Sea was almost devoid of ice three years later.

any conclusions about the solar constant, past or future, since the Solar Max program is so recent.

In the meantime, scientists are searching the record of ice cores for clues to past variations in solar activity. From old astronomical observations, they know that during part of the Little Ice Age the sun presented an unusually quiet face to the earth. From 1645 to 1715—a period called the Maunder minimum, for E. Walter Maunder, the 19th Century British astronomer who brought it to the attention of his contemporaries—observers of the sun noted that sunspots were almost totally absent. These spots, a regular solar feature, are dark areas where concentrations of the magnetic field reduce the flow of heat from the sun's interior to the surface. Normally, the number of sunspots varies on a more or less regular 11-year cycle. Why they disappeared during the Little Ice Age, and why they returned, is a mystery.

Whatever the explanation turns out to be, the chemistry of the snow that fell during the Maunder minimum was different. A team of French scientists analyzing a section of ice core from Antarctica's Dome C found that the snowfalls of the Maunder minimum were high in beryllium 10, an isotope formed in the upper atmosphere when it is bombarded by electrically charged particles emitted by the sun. Older ice layers dating from the Ice Age also have high concentrations of this isotope. Other chemicals appearing to rise and fall along with sunspot activity are nitrates, which, like the beryllium isotope, are formed by the interaction between charged particles and gases in the upper atmosphere. Scientists cannot yet say what the connections are between these isotopes, climate change and solar activity, but remain confident that the chemistry of ancient snow will eventually reveal at least some of the history of the sun.

Another extraterrestrial force has been proposed as a factor in the 180-year temperature cycles that Dansgaard found in the Camp Century core. Two Chinese scientists have advanced a novel and highly controversial hypothesis that links the O-18 oscillations with a celestial event that occurs every 178 to 182 years. Called a planetary synod, it is an astronomical arrangement in which all the planets except the earth are bunched together in a narrow arc of 90 degrees or less on one side of the sun. Ren Zhenqiu and Li Zhisen reviewed Chinese historical records for references to unusually cold periods and discovered that they coincided with planetary synods. One synod, for instance, occurred in 1665, just before the coldest decades of the Little Ice Age. (Unfortunately, the next synod, in about 1845, coincided with the warming trend that marked the end of the Little Ice Age.)

Ren and Li have speculated that the combined gravitational forces of the tightly grouped planets might alter the rate at which the earth travels different parts of its orbit. They think that when the earth is at its farthest from the synod—on the opposite side of the sun—its forward motion would be slowed slightly. As the earth nears the clustered planets, their gravitational tug would make it speed up. According to Ren and Li, if the earth were slowed down while winter reigns in the Northern Hemisphere, the winter would be lengthened, perhaps by three days or so, and summer would be shortened by the same amount. Such a shift would set the scene for a greater accumulation of snow and less summertime melting. Synods could make their effects felt over many years, since it takes the planets about two decades to move into a narrow arc and two more to disperse. A long series of years in which the lengths of the seasons shifted might tip the

Ramparts for a Sinking City

"There was the last night the greatest Tide that ever was remembered in England to have been in this River." So wrote the English diarist and naval expert Samuel Pepys after watching London's calamitous Thames River flood of December 7, 1663. Even the astute Pepys, who knew a record high-water mark when he saw one, could not have noticed that matters were getting worse; nor in his wildest dreams could he have guessed why.

Since the last ice age, the southern half of the land now called Britain has been steadily sinking as the earth's crust recovers from deformation caused by ice sheets. The process works in seesaw fashion (*below*). When glaciers advanced into northern Britain and depressed the land there, the ice-free land to the south was bulged upward by the displacement of the plastic mantle that lies under the more brittle crust. As the ice melted, the depressed crust and the mantle material started to return to their former positions. The resulting subsidence in the south, combined with the continued

rise in sea levels caused by melting ice sheets, makes the Thames increasingly vulnerable to a combination of weather conditions that tends to occur, on average, every hundred years or so.

When a large storm forms in the North Atlantic, a mound of water called a storm surge is created beneath the low-pressure system. Then, if the storm swings southward from the North Sea, part of the water rises even higher as it is funneled into the relatively narrow estuary of the Thames. The storm surge may then be pushed upriver by a strong following wind. If such a surge coincides with a spring high tide, the high water could inundate London.

To forestall such an event, British engineers have spanned the Thames River below London with the largest movable flood barrier in the world. Measuring some 570 yards from bank to bank, the billion-dollar barrier contains 10 gigantic gates that can be pivoted upward in approximately 30 minutes to dam the river and save London from a perilous legacy of the Ice Age.

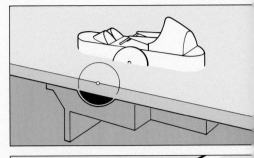

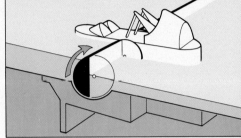

AN INGENIOUS PIVOTING DAM
Under normal conditions (*top*), a half-cylinder sector gate rests in a cradle on the bottom of the Thames River. But when flood waters threaten, the gate is pivoted upward hydraulically, forming a barrier 50 feet high. At right, one gate is held upside down on the surface awaiting installation in 1981.

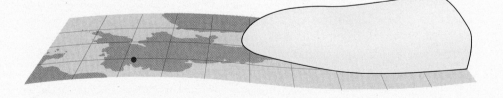

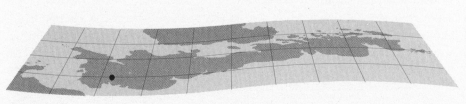

THE RISE AND FALL OF BRITAIN
About 18,000 years ago, the tremendous weight of mile-thick ice in the north of Britain depressed the earth's crust by as much as 1,500 feet. Nearly molten rock deep in the earth, squeezed southward, raised the crust under the rest of Britain by some 100 feet. When the ice melted, the earth's crust began to adjust the other way: Today the southern areas sink almost one foot each century.

delicate climatic balance just enough to bring an end to an interglacial.

Additional evidence of the abruptness of the transition from an interglacial to a glacial climate has been found in a peat bog in Alsace, in northeastern France. The bog has remained undisturbed for 140,000 years—a span that includes all of the last interglacial, the Ice Age and the present interglacial. Genevieve Woillard, a Belgian botanist, examined the pollen in the layers of peat formed some 115,000 years ago, in the final three centuries of the last interglacial. In the oldest layers she found the pollen of trees that flourish in a temperate climate, with firs, oaks, alders and hornbeams especially plentiful. During the next 125 years or so, spruces, which are cooler-climate trees, gradually gained ground over the temperate-forest species and became dominant. In the century that followed, the climate cooled further,

The sun's most active regions are starkly outlined in this computer-enhanced image of X-ray emission taken in 1973 from the Skylab satellite. Some scientists believe that events in these regions—including sunspots and solar flares—induce climate changes on earth.

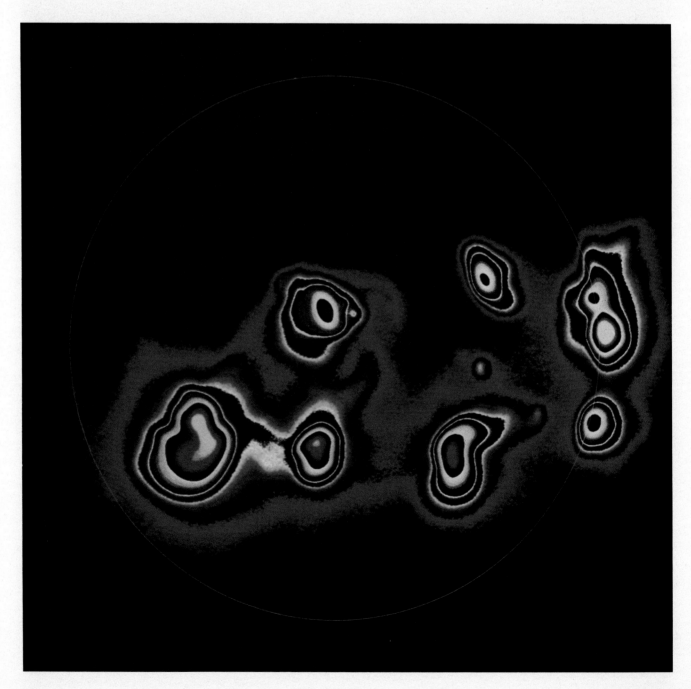

the temperate-forest trees became less numerous, and pines began to grow alongside the spruces.

Then, in a very short period—perhaps no more than 20 years—there was a radical change in the vegetation. The temperate-climate trees disappeared altogether, along with companion plants such as mistletoe, which requires summer temperatures higher than 60° F. to survive, and English ivy, which cannot endure winters during which the temperature stays below 30° F. for long periods of time. At the end of the rapid transition, the forest was very much like that of modern-day northern Scandinavia, which lies some 1,400 miles north of the site of the Alsatian bog.

Woillard pointed out that the early stages of vegetation change, when temperate trees first began to dwindle in number, would probably pass unnoticed today; forests are now carefully managed—at least in Europe and some other developed regions of the temperate zone—and can no longer be considered natural plant communities. According to Woillard, an undetected and relatively fast "northernization" of forests, like the one that took place at the onset of the Ice Age, could be imminent.

Other kinds of human activity have made the earth a very different place from what it was during the last interglacial, and some of the changes might hasten or retard the arrival of any future ice age. A great deal of scientific attention is being focused on the effects of the carbon dioxide added to the atmosphere, principally by the burning of fossil fuels. Increased carbon dioxide levels might possibly boost the heat-trapping greenhouse effect enough to override, partially or completely, any natural tendency toward lower temperatures.

On the other hand, the ways in which civilization has altered the earth's landscape may have the effect of encouraging heat loss. George J. Kukla and Jeffrey A. Brown of the Lamont-Doherty Geological Observatory have examined satellite photographs made of the winter snow cover across the North American continent and the Soviet Union. Areas where forests have been cleared for agriculture have a high albedo: Snowy fields can reflect as much as 70 per cent of the sunlight reaching them, while dense forests reflect only 10 to 20 per cent. The vast areas of brightly reflective farmland in the photographs Kukla and Brown studied prompted them to wonder whether the large-scale felling of forests that took place in Europe not long before the Little Ice Age made the climatic deterioration more severe than it otherwise would have been. If artificially created areas of high albedo caused atmospheric cooling then, there is no reason to think the same effect would not operate in the 20th Century.

Modern civilization has injected its own brand of change into the network of climate systems that respond to changes in the earth's relationship to the sun. This makes even more difficult the task of the scientists who are seeking to compare the past with the present in order to predict the future. They are grappling with a multidimensional puzzle whose pieces are constantly being redefined. Long ago, in 1837, Louis Agassiz framed the nature and direction of ice age studies while he was announcing his revolutionary ideas on glaciers to the Swiss Society of Natural Sciences. Scientists of every discipline who wish to understand the ice ages, he said, must be guided by "the idea of a progressive development in all living beings, of a metamorphosis through several states dependent on one another—the idea of a comprehensible creation." Ω